The Best of Technology Writing 2007

DIGITALCULTUreBOOKS is a collaborative imprint of the University of Michigan Press and the University of Michigan Library dedicated to publishing innovative work about the social, cultural, and political impact of new media.

Steven Levy, Editor

The Best of Technology Writing 2007

THE UNIVERSITY OF MICHIGAN PRESS AND
THE UNIVERSITY OF MICHIGAN LIBRARY
Ann Arbor

Copyright © by the University of Michigan 2007
All rights reserved
Published in the United States of America by
The University of Michigan Press and
The University of Michigan Library
Manufactured in the United States of America
⊚ Printed on acid-free paper

2001o   2009   2008   2007   4   3   2   1

A CIP catalog record for this book is available from the British Library.

Library of Congress Cataloging-in-Publication Data

The best of technology writing 2007 / Steven Levy, Editor.
    p.   cm.
   ISBN-13: 978-0-472-03266-2 (pbk. : alk. paper)
   ISBN-10: 0-472-03266-6 (pbk. : alk. paper)
   1. Technical writing.   I. Levy, Steven.

  T11.B458   2007
  600—dc22   2007023386

Contents

Introduction

Not long ago, in one of my periodic cleaning binges, I came across a steno notebook dated "1984 Northwest." The pen scratches within provide a record of my journey to Seattle and environs in the middle part of that year, now (gulp) over two decades past. It was illuminating, and a little jarring for me, to recollect the impetus of that trip, what I found, and especially how I went about collecting information.

In those distant days (9,000 years ago in Internet time), I was writing a column for a magazine, now long buried, called *Popular Computing*. Unlike our sister publication, *Byte,* which was the unofficial geek bible (published monthly, at a bulk that made *Vogue* look pencil thin), *PopCo* attempted to reach a less technical audience who had recently stepped into the world of personal computers and software and were both fascinated and a bit dazed as a result. In those days, of course, anyone without an MIT degree who routinely used a computer—be it an IBM PC, an Apple II, or a Kaypro—was viewed as an adventurer in a strange realm. Simply using a word processor was a sort of desktop X-game event. At one point our magazine ran a story profiling women who used personal computers—not Silicon Valley executives, not IT consultants, not female programmers, just those then-rare creatures who had mastered the command line without benefit of Y chromosomes. (I am not making this up.) Our crew at *Popular Computing* was supposed to demystify this brave new world by providing sound advice about how to buy and operate these wonderful but often confounding new machines. We also strived to recount some of the stories and reveal a bit of the sociology behind this new phenomenon. The latter was the point of my column, which ran every month at about 2,500 words, a length exceeding the average cover story of my current employer, *Newsweek*.

I wound up at *Popular Computing* because two years earlier, in 1982, while working as a general magazine freelancer, I had stumbled on the subject of technology and become hooked. The big draw for me was speaking to incredibly smart people who actually gave candid interviews. After spending the early portion of my career talking to pop musicians, sports stars, criminals, and politicians, these computer whizzes were a welcome tonic. Also, the field was pretty much virgin territory for a professional writer. Computer journalism was just beginning to switch from hobbyist publications to slick operations like mine, which was owned by McGraw-Hill. I could gorge on low-hanging fruit.

By 1984, I'd certainly heard a lot about Microsoft, and I had even interviewed its young CEO in New York City the year before. But I wanted to learn more. I'd also noticed that Microsoft was now only one of many interesting tech destinations in the Northwest—and I'd never been to Seattle. So I packed up and went. On very little notice I managed to book some great interviews. I talked to Tim Patterson (author of QDOS, the operating system he sold to Microsoft that became the basis of its personal computer empire); Bob Wallace (now sadly gone), one of the first Softies, who was then pioneering the concept of shareware; and Rudi Deitzmann, author of Panorama, a great database that ran on the brand new Macintosh computer (I took a ferry to Orcas Island to speak to him). Still, the centerpiece of the trip was my time at Microsoft, then a smallish but ambitious company headquartered in one large building in Kirkland, across Lake Washington from Seattle. A single phone call to the Microsoft public relations company in Portland secured me a full day's worth of touring and interviewing. I spoke to all the top people, including a fiery young executive named Steve Ballmer, before being unceremoniously ushered into Bill Gates's office for an open-ended chat. (It was less an appointment than an "Oh, Bill has some time now" kind of thing.) I remember Gates sprawled on a couch in his office, hands behind his head, free-associating his way through issues of industry, geography, and geekdom—a veritable Lord Buckley of the software world. Our conversation must have lasted two hours, without any nervous couriers peeking in to urge Boy Wonder on to his next appointment.

In the mid-1980s technology journalism was a specialty act. Editors knew that something important was going on, but they also feared that readers would be turned off because the subject was, well, *computers*. Those were the days when most red-blooded Americans felt proud to declare they didn't know a fig about computers—even CEOs of huge companies took pride in keeping their offices clean of PCs. Most major magazines and newspapers thought it was important to have someone on the staff who knew enough to cover what was becoming a big story in economics, science, and even culture. But if you wrote for a general publication, almost every paragraph required at least one or two detailed sidebars explaining some bit of technical jargon.

I exhume the gestalt of those hazy days past to contrast them to our current era, one noted for saturating us with information about the digital revolution. The terms that were weird then are now taken for granted. Computers are part of everyone's life, and it's not unusual to discuss them with an ardor similar to that of car aficionados comparing their rides. Meanwhile, the Internet has become so deeply woven into our lives that we naturally tend to obsess about it as well. Everyone has an opinion because even your grandparents are likely to use it. It's a national treasure and an all-purpose scapegoat, a disrupter and a builder, a cesspool harboring porn and fraudsters, and a utopian haven for free markets and free speech. People keep scorecards on new startups and are routinely dazzled by the money that befalls their founders. Lottery jackpots are rounding errors compared to some of these riches.

Now dozens of publications have Silicon Valley bureaus packed with reporters who are trained to root out every last detail of every technological development. The major figures in tech are rock stars, only much bigger than real rock stars (a species laid low by the Long Tail). Bill Gates, who cheerfully hung out with a visitor in 1984, has his time parceled out more penuriously than the President. Most journalists, in fact, would find it easier to nail a White House interview than a pizza joint sit-down with Apple CEO Steve Jobs, who ordered olives on his pie when we dined in 1983 and discussed his weird new computer with a name that seemed doomed to be forever confused with a certain confection sold at McDonald's. And if you do score an interview with one of the big guys, depend on being accompanied by at least one PR person feverishly taking notes and monitoring the conversation to see if you venture out of bounds. In fact, depend on the same monitoring process not only when you interview someone in those big companies but when you're meeting with a founder of a startup still in beta as well.

In short, the low-hanging fruit has long been picked. With hundreds of journalists scrambling for inside angles on every story, with sources coached on what they can or can't say, and buzzards from the Blogosphere picking clean the bones, it's harder than ever to create fresh, memorable accounts that stand out from the din.

And yet, every year some writers manage to do just that. This book is proof.

In helping select the choices for the second edition of *The Best of Technology Writing,* I couldn't help but be impressed by the persistence and skill of our authors. Their work splendidly contradicts the extremists in the broadband peanut gallery who insist that this form of expression—specifically long-form, well-researched journalism—is a doomed artifact of a predigital era. We are supposed to think that such efforts can be replaced by "Wisdom of the Crowd" enterprises, such as Google and Wikipedia, and a cumulative "river of news" that will stream past our screens. Not so fast, folks. Collective enterprises that draw on mass participation may enrich our lives, but there are some things that a collective approach can't duplicate—like the tech stories we killed all those trees to reprint here. Crowds may be wise, but they are not witty. Great stories are reported and written by individuals, using *selective* intelligence. It may well be, as Kevin Kelly wrote in an article that had John Updike gasping for breath, that books themselves might become fodder for a vast collaborative masterwork. But only Kevin Kelly—a deadpan iconoclast who's also a celebrated author himself—could have written that article, with its unique blend of provocation and vision.

This is not to say that our collection includes only the efforts of mainstream media. Justin McElroy's piece about how a game studio might save a depressed small town in Ohio comes from the *Escapist,* a weekly online publication about gaming culture that I hadn't heard of before helping with the collection; now I follow it avidly. Another author, Kiera Butler, is a recent journalism school graduate precociously hitting a home run in one of her first at bats. Jaron Lanier's jeremiad "Digital Maoism" appeared in the blog-like online science compendium Edge.org. It is nonetheless true that some blog writing doesn't translate well to the medium of print. And so the excellent ongoing work in places like Engadget, Scripting News, TechCrunch, and other real-time dispatches doesn't appear here. But we were able to identify some blog postings that survive the transition—ones where the welcoming platform of the Internet drew out the inner journalist in those so-called outsiders. John Gruber's contribution comes from Daring Fireball, where one finds some of the most insightful writing about Apple. Aaron Swartz began writing his blog while a Stanford freshman and continued it after he dropped out, joined a startup, and got rich.

Still, as my predecessor, Brendan Koerner, pointed out in his introduction to last year's collection, the best tech writing is still found where'd you most expect it: top-notch publications that seek out the best writers. So it's no surprise to find pieces that ran in *Wired, Atlantic Monthly, Fortune Magazine,* the *New Yorker,* and the *New York Times Magazine,* as well as online ventures that seek out the cream of the writing crop, such as *Salon* and *Slate.* Hundreds of bloggers provided dispatches on the "net neutrality" debate, a somewhat arcane controversy that, improbably, was Net Nation's number one public policy contretemps in 2006. But it took a careful reporter like Farhad Manjoo to wade through the rhetoric and carefully explain its implications in a manner that turns casual readers into experts—and without dishing out the writerly equivalent of Ambien in the process. Clive Thompson's superbly researched and rendered piece on

Gordon Bell's fascinating experiment in constructing a digital memory bank ran in *Fast Company;* because we didn't want to include more than one piece by any author, it was a toss-up between that and his deep dive into Google's adventures in Chinese censorship that ran in the Sunday *New York Times Magazine.*

In 1984, many media types would have scoffed at the idea that technology is the major story of our times, the one that our era will be remembered for. Now, when it is clear that the modern computer has had a profound impact on virtually every aspect of our lives, that sentiment seems about right. Fortunately, technology writing has risen to the task of creating a worthy record of this historic period. And, insofar as 2006 goes, you're holding that record in your hands. It might be harder to pin down Bill Gates for an interview these days, but, as this collection demonstrates, great stories are being written. And it's more important than ever to read them.

Kevin Berger

The Artist as Mad Scientist

She is an intellectual and emotional storm. Her renowned public artworks are reshaping the ways we think about science. Activist, environmentalist, and former rock promoter Natalie Jeremijenko turns the art world upside down.

It's an icy spring morning, and Natalie Jeremijenko skates into the Soy Luck Club on Rollerblades. The boutique cafe in New York's West Village has polished concrete floors, brick red walls, and burnished wood counters. It feels like it was designed to be featured in one of those big modernist architecture magazines. But Jeremijenko's chaotic energy seems to melt the frosty interior into thin air.

Wearing a parka with a fake fur collar over a tight dress, she rolls around the glossy Knoll furniture, talking nonstop about her latest art project. She has all kinds of science degrees and drops terms in biology and mechanical engineering the way most people do the names of film stars. With her animated eyes and sly smile, her blond hair pulled loosely behind her head, she has the magnetism of a natural actor. Yet her enthusiasm often overtakes her logic, and her sentences dart around like children. Born and raised in Australia, she retains her Aussie accent and seems to live so comfortably on abstract planes that at times you don't know where she's coming from. And that, her husband, Dalton Conley, a New York University sociology professor and writer, later says, goes for him too.

Alighting at a table, Jeremijenko, 39, explains that her work is "all about creating interfaces that draw people into the environment and get them to reimagine collective action." She cracks open her laptop and displays an image of 100 polycarbonate tubes, or "buoys," that she's engineered to glow when fish swim through them in the Hudson River. Yes, she really has government approval to position the buoys in the river. Given her day job as a professor, she convinced state environmental officials her project was all about science. But never mind that. Did you know the fish were on Zoloft? All the antidepressants that New Yorkers take are flushed through their urine into sewage treatment plants, which overflow into the river. You doubt her? Go to the Whitney Museum and see one of her drawings hanging on a wall by a bathroom. It features a woman's bottom, her pants below her knees, on a toilet seat. It asks, "Why are the Hudson River fish and frogs on antidepressants?" Printed on it in tiny letters are actual studies that attest to the chemical drug compounds in the waterway consumed by the unsuspecting bass, sturgeon, and crabs.

Anyway, when the buoys light up, you can feed the fish food treated with chelating agents to help cleanse the PCBs from their blood, planted there from decades of General Electric dumping waste into the river. The fish food, in fact, will not be much different from the energy bars we're always eating on hiking trails. "The idea that we eat the same stuff is a visceral demonstration that we live in the same system," Jeremijenko says. "Eating together is the most intimate form of kinship. By scripting a work where we share the same kind of food with fish, I'm scripting our interrelationship with them."

Oh, and one more thing. Do you know about the American doctrine that says a corporation has the status of a person and enjoys all the legal protections afforded by the constitution, including the right to own property? Well, beginning this week, Jeremijenko is selling the buoys to collectors. With the money, she plans to form a corporation called Ooz Inc.—*zoo* spelled backward—and put the fish on the board. That way the fish, as shareholders, will acquire personhood and have a say in the preservation of their grungy habitat.

Is she kidding? No, she's not. She wants us to feel as connected to wildlife in New York City as we do in the Adirondack Mountains. And reflect on the ways we impact nature and the ways it affects us. She's a maverick environmentalist whose field notes are public artworks. But she is being playful, a hallmark of her art and personality and the trait that allows her work to stand out in the vital cultural arena where art and science collide.

The thesis of the famous 1959 essay *The Two Cultures,* by British scientist and novelist C. P. Snow, that science and the humanities represent two worlds that don't meet, has vanished into history. Science has become a daily topic in the cultural conversation. Biologists like Richard Dawkins and physicists like Brian Greene write so fluently about their fields that readers pore over their books with the passion of pursuing a good story. And ever since Oliver Sacks set fountain pen to paper, everybody at cocktail parties seems to be an expert in neuroscience.

Science is now permanently stitched into the arts themselves. Filmmakers have long flocked to science for Frankenstein themes about controlling nature. In recent years, genetics, math, and physics have informed, respectively, genuinely moving works in literature (Richard Powers's *The Gold Bug Variations*), theater (David Auburn's *Proof*), and opera (John Adams's *Doctor Atomic*). But it's the artists working in studios, labs, and garages out of pop culture's shadow who have been melding science and humanity in the most challenging, fascinating, and profound ways. And that goes for Jeremijenko.

She set her artistic course in 1994 as a consultant research scientist at Xerox Palo Alto Research Center, the famed Silicon Valley lab where artists and computer scientists are paid to let their minds wander. There, she created *Live Wire,* a vibrating cable that symbolized the energy consumed by the Internet. Just as the Web was being celebrated as a virtual reality where egalitarian dreams came true, Jeremijenko was saying the electricity required to run it and the people being used to build it were creating environmental and labor conditions that were far from utopian.

Her next project, *Suicide Box,* was also a grim twist on cultural infatuation. As people seemed obsessed with NASDAQ and Dow Jones numbers, Jeremijenko created a numerical calculation for suicide. Using a motion-sensitive camera to track vertical movement off the Golden Gate Bridge for three months, she recorded 17 people who appeared to have leaped to their deaths. In 2004, she aired out her politics. During the Republican convention in New York, she fashioned medical face masks for bicyclists to wear around the city. The grime the riders inhaled was displayed in a sooty bar beneath the words *Clear Skies,* an ironic echo of the Bush administration's air pollution policy. On the lighter side, her proposed project for parking lots, in which cars are assigned spaces by color to create artistic patterns in the lots, is a quite lovely idea.

While public spaces are her main stages, her works have

been exhibited in art museums from Sydney to Amsterdam to San Francisco. Earlier this year, along with her drawing, her outdoor installation *For the Birds,* a wry comment on avian flu, was featured in the 2006 Whitney Biennial, the always controversial assembly of contemporary art at the New York museum. She has held academic positions at NYU and Yale and currently is an assistant professor in visual arts at the University of California at San Diego. True to her frenetic life, she commutes to San Diego two days a week from her home in New York.

Art critics at major newspapers seldom pay attention to Jeremijenko, busy as they are with reviews of the latest Monet exhibit or Henry Moore retrospective. "The art world is a very prissy little thing over in the corner, while the major cultural forces are being determined by techno-science," Jeremijenko said in 2000. She said that in the *New York Times Magazine,* so, yes, she's gotten her fair share of feature coverage. Accolades for her work are easily found in avant-garde art magazines and Web sites, science and design journals. In 1999, MIT's *Technology Review* named her one of the country's top young innovators, and in 2005, *I.D. (International Design)* magazine named her one of the 40 biggest "influencers" in architecture and design, listing her alongside Steve Jobs, Frank Gehry, and Rem Koolhaas.

"Oh, yes, Natalie is a star in the alternative art world," says Stephen Wilson with a laugh. Wilson is a multimedia artist and art professor at San Francisco State University and the author of *Information Arts,* the definitive tome of contemporary artists at work in the fields of science and technology, which includes numerous references to Jeremijenko. Philippe Vergne, who cocurated the 2006 Whitney Biennial, also chuckles when he mentions Jeremijenko, recalling the first time he met her, six years ago. Then, Jeremijenko had just finished hanging six maple trees upside down at the Massachusetts Museum of Contemporary Art. She wanted to upend the view of city trees as pretty little rows on streets. "Honestly," says Vergne, "I thought she was totally mad." Which only kept him coming back to her work. "From science to politics to genetics, she's really putting her finger on controversial questions that are framing our culture right now."

To be sure, she's not alone. As Wilson's book (900 pages!) makes clear, Jeremijenko is only one player sounding the noisy times with notes from biology, engineering, zoology, genetics, and physics. Christa Sommerer and Laurent Mignonneau illuminate the harmonies between real and artificial life, and the human impact on ecosystems, in wondrous interactive 3-D exhibits that simulate plants and animals. With wit and strange beauty, Alexis Rockman unearths urban wildlife—seagulls, rats, cockroaches—in fantastical paintings that the late biologist Stephen Jay Gould has praised for splintering calcified scientific views. And the ways technology alters our view of ourselves and the environment, seen through ingenious robotic installations, inform the work of Ken Goldberg and, more apocalyptically, that of Eduardo Kac.

Kac is the movement's most notorious star. In 2000, he made headlines when he injected a jellyfish gene into an albino rabbit, causing it to glow green when illuminated. The idea was to stimulate discussion about the ethics of transgenics, although Jeremijenko calls it "a stupid piece that is exactly not what bioart is." The grandstanding exhibit skirted the issue that sparks Jeremijenko and her comrades in political troublemaking, the Critical Art Ensemble, whose radical adventures in public art have earned international renown. To them, the burning issue in science is how government scientists and biotech companies are shaping how we think about genetics, and how gene mixing is employed in the manufacture of food and medicine, despite questions about its ecological effects.

Jeremijenko spliced into the controversy in 1999 with her project *One Trees*. When daily news about decoding the human genome had us all fearing we were programmed by DNA, she and a California nursery produced hundreds of clones of a walnut tree from its stem-cell-like tissue. She placed the tiny sprouts in individually sealed cups and displayed them at Yerba Buena Center for the Arts in San Francisco. The various shapes into which the plantlets sprouted in uniform environments underscored that genes alone don't sit in nature's director chair but are one of many biological processes. Later, she planted 20 pairs of the trees in various places in the San Francisco Bay Area. Now she gave urban reality its turn in the spotlight by demonstrating how social conditions—the trees were planted in poor and wealthy neighborhoods—caused the genetically identical trees to blossom into a bounty of sizes, some rising into the open sun with vigor, others drooping under the industrial shade.

"*One Trees* is one of the landmarks so far in art and science," says Wilson. "A lot of artists just sit back and comment on the world. Natalie actually went out and employed the science. It was very powerful, more powerful than just sitting back and commenting."

Jeremijenko lives her work like few artists. She even tries to involve her three kids, much to the chagrin of Conley, who laments that family vacations to Costa Rica and Australia turn into fact-finding missions for her ongoing project about manufacturing, "How Stuff Is Made." Finding the woman behind the artist, though, can be exhausting. Ask her what's personal about her art and she will say, "My thesis is nothing can't be autobiographical. The idea that there is a rational truth out there that is not embodied in a person's politics is something I can't understand or subscribe to."

But ask Conley and he will tell you her "work is very personal, which doesn't always come through because she presents a scientific front. But the initial spark is always personal." Spend enough time with Jeremijenko and the woman in her art does come into view. You see an artist with a frightening drive, dark, passionate, and joyful motivations. It's just that, like her art, she requires translation.

People are beginning to fill up the Whitney Museum on a Saturday morning. Some trickle down to the gift store on the lower floor, where Jeremijenko cuts a path through them and rolls aside a bookshelf on wheels. Behind the shelf, she retrieves the laptop that controls her installation, *For the Birds,* created with Phil Taylor and her artist collective, Bureau of Inverse Technology, set up on a patio outside the store. Museumgoers are not sure what Jeremijenko—wearing a long silver coat, black knee boots, and straw cowboy hat—is up to. But she pays them little mind. She is focused on showing me how *For the Birds* works. As I can see, there are translucent bird perches affixed to the outdoor patio walls. When real birds land on them, they set off prerecorded voices that warn museumgoers about the encroachment of avian flu. Next to the perches are miniature reproductions of some of the paintings inside the museum. "It's the Whitney Biennial for the birds," Jeremijenko says with a short laugh.

Before she goes into detail about *For the Birds,* a companion to her buoys for the Hudson River, Jeremijenko regales me with the philosophy behind them. She calls both projects "Hudson River School 2.0," a reference to the nineteenth-century painters who lushly rendered the eastern countryside a pastoral Eden. "The Hudson River School romanticized the American landscape for the first time," she says. "It showed it as beautiful vistas of great green expanses with great shimmering blue skies and lots of water. It was a view of nature as something out there, something pretty, something apart from people." Although the Hudson River School may have enchanted people for generations, drawing them closer to nature, its views are now quaint and counter-productive. The American landscape after two centuries of human development is not a pretty picture. It's time for environmentalists to stop seeing nature as scenery that should be preserved like a painting in a museum, Jeremijenko says, and more like a dying body that needs to be nourished back to health.

With Hudson River 2.0, she continues, we get "a view of nature where we're inside it, interacting with it, where urban forms are a part of nature and act as their own natural systems." It's a global view, informed by ecology, where city and country, humans and birds, exist in an interrelated dance. And right now, the dance of nature is a dangerous waltz with a deadly flu virus in the mix. Or so the birds in Jeremijenko's Whitney Museum piece are trying to tell us. Or would, if New York's real birds would cooperate by landing on the perches and triggering the voices. Given their reluctance at the moment, Jeremijenko, sitting on a window ledge in the gift store, her computer in her lap, has to intervene and manually play the bird voices for me.

"Tick, tick, tick. That's the sound of genetic mutations, of the avian flu becoming a deadly human flu," says a professorial male voice. "Do you know what slows it down? Healthy subpopulations of birds. Increasing biodiversity, generally. It is in your interest that I'm healthy, happy, well fed. Hence, you could share some of your nutritional resources instead of monopolizing them. That is, share your lunch."

Next comes a female voice. "You have such a strange

relationship to ownership that holds across species. I'd like to suggest that we share the land and its productive capacity—the worms, the plants, the future generations of seeds, the nesting grounds. Do you think you own this too?" The haughty voice continues. "You know those mute swans now dying all over Europe? They don't normally migrate," she says. When it comes to bird flu and human deaths, "You're bringing it on yourselves. But that means you can fix it. The first step is to give me a little bit of that bar."

These are some brainy birds. They're telling us how the destruction of biological diversity is a crime against nature and increases the risk of disease. Jeremijenko explains that wild birds in Europe and Asia, fleeing ailing wetlands, are forced to roost near scummy ponds on farmlands, where they come in contact with infected chickens. Yet rather than preserving wild lands, she laments, the international response has been the "mass slaughter of millions of birds," which only fans the flames of the flu.

"The birds are arguing that the reason we have diversity in nature is to protect us against disease," she says. "The birds are arguing that if we were to address the problem effectively, with a systems-level view, we would increase the health of domestic and wild birds, and that would be our best protection." Her birds, she says, also remind us we don't live in plastic bubbles. "The greatest vectors of bird flu have been freeways, airports, and railways. People get on with infected birds, get off, and trade at stops along the way. It's human migration that is transmitting this disease, not the migration of wild birds themselves."

Jeremijenko always sounds like an excitable activist. But she does do her homework. A recent report by the United Nations Environment Programme concluded, "Restoring tens of thousands of lost and degraded wetlands could go a long way towards reducing the threat of avian flu pandemics." Ecologists at the University of Georgia, as reported by *New Scientist,* "have shown that killing wild animals with a disease like flu could actually lead to more infected animals, not fewer." The theory is that older animals build up immunities to disease and so killing them leaves the younger and more populous ones vulnerable. As for roads and railways leading to the outbreaks of bird flu, Nial Moores, director of the conservation group Birds Korea, says, "There is abundant evidence that poultry flu is spread over significant distance through the transport of poultry."

Jeremijenko's views about bird flu have also been informed by one of her brothers, Andrew, a physician, an epidemiologist, and the head of influenza studies at a U.S. Navy medical research facility in Indonesia, where the first human cases of bird flu were studied in detail. The naval medical center, which *Scientific American* called "the largest, most experienced and best-equipped avian influenza laboratory in the country," was shut down earlier this year due to ugly politics inside the Indonesian government, related to lingering resentment against the United States for backing the independence of East Timor, formerly controlled by Indonesia, in 1999.

The mere sight of Jeremijenko's bird perches doesn't exactly engage museumgoers in the important political and ecological debates over bird flu or enrich their reveries with a theory of natural systems. In a previous project, "Feral Robotic Dogs," Jeremijenko and her students hacked toy robot dogs, implanted chemical sensors in their noses, and unleashed them in urban "brown sites," public spaces where toxins like lead and arsenic, remnants of chemical plants, lurked beneath the surface. With the media in tow, and a bunch of happy schoolkids, Jeremijenko set the robotic dogs loose alongside the Bronx River, near a long-shuttered Con Edison plant. It created just the kind of spectacle she loves. I remark that the bird installation seems less successful and represents a significant problem with conceptual art—that its effects often remain locked in the artist's head. Jeremijenko bristles.

"In terms of conceptual art, I'm a populist," she responds, perfectly serious. "My strategies of representation are supposed to be questions. What are those perches? They're not supposed to be hammering you on the head as some kind of didactic science lesson."

Well, she has to admit that a lesson given by birds on the evils of land ownership is didactic, if not anthropomorphic. "There's nothing wrong with being anthropomorphic," she says, sounding as peeved as one of her talking birds. "That's how we understand the world. Contemporary science doesn't pretend we're not human and have a solely objective view of things. We only have certain tools to understand the world. So I'm not scared of anthropomorphism at all. When science becomes unbelievable is when it pretends it's coming from nowhere, that it's a universal description of everything. That doesn't make sense anymore." And too often in our culture, she says, when empirical science rules, activism fails. "I don't think there's a scientist anywhere in the world, in any discipline, that has the kind of power to make as much of a difference as an interested, engaged, diverse community."

She pauses and grins, her wry humor having returned. "Besides, walk through the rest of the Whitney and ask me again if I'm not a populist."

She does have a point. Strolling through the museum, we stop by two large holes knocked out of a drywall. Through the ragged sheetrock openings you see horizontal tree branches hanging from the ceiling by chains. The branches rotate in a circle and drip wax from their tips onto the floor. I'm not certain why the piece is called *Intelligence of Flowers.* I comment to Jeremijenko that I think it suggests how art breaks through the "fourth wall" of museums. "Radical, huh?" she says dryly.

We pass by a tangled mass of bicycle handlebars sheathed in red, white, and blue. It appears that a massive bicycle troop crashed into a wall at the end of a Fourth of July parade. Even Jeremijenko is not sure what to make of the wreck of aluminum tubes. However, she does nod approvingly at a lifelike rubber toddler kneeling in the middle of the museum floor with a green parka over its head.

Abstract art apparently has the opposite effect on Jeremijenko as it does on most people. Triangle drawings, green bottles, and huge paintings of red high-heel shoes draw out her more personable side. As we walk through the fluorescent pawn shop of art, she talks about her long haul in academia. Showing an aptitude for math and science, apparently a rarity for girls in Australia, she was "pipelined" into science studies and never got out. In Australian universities, she earned bachelor's degrees in biochemistry and physics and a master's in English, and she completed the research but not her PhD dissertation in neuroscience. She transferred to Stanford and did all the course work but didn't finish her dissertation for a PhD in mechanical engineering. "I'm not that interested in credentials," she says coyly. Finally, she returned to Australia and completed her PhD in computer science and electrical engineering at the University of Queensland.

Not that school was enough, however. In 1987, she had a daughter with her high school sweetheart, a rock musician, and named her Mr. Jamba-djang Vladimir Ulysses Hope. She was reading and loving Nabokov and Joyce at the time, she says, "and got a little carried away."

In 1988, she got involved in another all-consuming project. With a buddy, Peter Walsh, she borrowed $5,000, ostensibly for a car loan, and started a rock festival called Livid. They wanted to spotlight Brisbane's rock scene, notably the Go-Betweens, and surround the musical stage with theater and visual art, an idea that predated the Lollapalooza Festival by three years. As usual, she was doing 10 things at once. "I remember one day, while doing neuroscience research, I was slicing up rat brain in the lab and negotiating with Nick Cave on the phone. My advisor was watching me and said, 'Natalie, take a break and get this arts thing out of your system.'"

She never did. In fact, the Livid Festival sowed the seed of public art in her. "When you have to produce art that works with 20,000 inebriated 20-year-olds, it becomes very different than doing reverent, hushed, respectful art," she says. "One year, we worked with Act Up to produce huge frozen ice penises that, in the middle of the Brisbane summer, everybody was licking. We had helicopters fly over and drop cocktail umbrellas printed with political messages on the crowd. We soaked rope in gasoline, rolled it in gunpowder, and placed it on the grounds. If you dropped a cigarette, it would light the rope and a flame would sear through the crowd. It was a tremendous spectacle. So with the festival, I really learned how to create culturally exciting art and different forms outside of the museum."

We leave the Whitney and take a cab to the West Village. Along the way, Jeremijenko insists the driver drop us off on the West Side Highway, an illegal move if ever I've seen one. Near a dilapidated pier on the river, she points out where she plans to put her fish buoys in the water. After a brief oration on the industrial history of the area, we walk to a restaurant called the Spotted Pig, across the street from her apartment. She quizzes the waiter on whether the pinto beans are cooked in animal fat, and when she's reassured they're not, she orders them, a salad, and a glass of wine.

In a buoyant mood, she explains that she grew up in tropical Mackay, a coastal city alongside the Great Barrier Reef. Her father is a family physician, and her mother, who has a PhD in education, is a schoolteacher. The first thing she tells me about her mother, with affection, is "she was the first person in Australia to have a microwave. She was a great believer in domestic technology—it enabled her." Her parents were also great believers in kids; they had 10 of them. Natalie is the second oldest.

The five girls and five boys are "an interesting bunch," Jeremijenko says with understatement, offering me some examples. Her oldest sister, Valerie, is the dean of students for a branch of the Virginia Commonwealth University, based in Qatar. Her brother Michael was a commercial pilot until he decided to go into coal mining in New South Wales with their brother Paul, who tired of his jobs as a professional football player and physiotherapist. Peter is a stuntman who has performed in countless movies, including, most recently, *The Matrix Revolutions* and *The Matrix Reloaded*. And her sister Melanie is a math and science teacher in Darwin, Australia, where "she works on the front lines of extraordinary racial and class violence."

Seizing the rare moment when she talks about herself without using the word *hermeneutics,* I ask her about Jamba, her now 19-year-old daughter. She sighs. When Jeremijenko left Australia for the United States in 1994, it wasn't to work on her art at Xerox Parc. She went looking for her daughter.

Jeremijenko broke up with her boyfriend when Jamba was a toddler, leaving her a single mother. When Jamba was four, her father wanted to spend time alone with her. In a decision that continues to haunt Jeremijenko, she agreed. She packed a bag for Jamba, but her daughter never came home. Her ex-boyfriend disappeared with Jamba for a year and a half. Jeremijenko desperately tried to find them. Finally, she heard from a contact, a flight attendant, who told her that she had seen Jamba and her father on a flight to California. Indeed, the father had moved in with a new girlfriend in San Francisco. Distraught and relieved at the same time, Jeremijenko packed up her studies and moved to Berkeley, where she lived with a friend, and later by herself in San Francisco.

For years, Jeremijenko was trapped in a Bleak House of California and international custody law. Sorting out which foreign parent deserved U.S. custody of a child born in Australia was "a vicious and ugly carnival," she says. And one that marched against her. Judges sided with her ex-boyfriend, and Jeremijenko was granted only visitation rights until Jamba was in high school, when her daugher finally moved in with her. Later, she would tell me that she couldn't do anything right in court. "If I showed emotion, I was hysterical, if I didn't, I was cold and uncaring, disinterested. It was a trial of motherhood and made me understand how women could be burnt as witches."

It was during those first years in the Bay Area that Jeremijenko created her project *Suicide Box*. Was it her response to depression over her daughter? "Fair enough," she says, taking a deep breath. A few moments of silence pass. "I hope you're not going to depict me as a suicidal maniac."

Changing the subject in the restaurant, Jeremijenko asks me if I've read Conley's books on families and race and his memoir called *Honky*, about being one of the few white kids to grow up in a housing project on the Lower East Side in the '70s. When I say no, she calls him on her cell phone and tells him, considering they live across the street, to bring his books down to the Spotted Pig. With the trim body and no-nonsense stare of a lightweight boxer, Conley looks uncomfortable in the trendy restaurant. Similarly, he speaks in a native New York accent seldom heard in the upscale Village. The hardcore New Yorker certainly seems the polar opposite of his wildly cerebral Australian wife. But he does share her wry sense of humor. "I've come to rescue you from Natalie," he says, handing his books to me. Jeremijenko smiles, sort of.

A week later, Conley, 37, is sitting in a coffee-and-muffins cafe down the hall from his NYU faculty office, located in a recently refurbished building in SoHo. He's director of the university's Center for Advanced Social Science Research. As candid as his wife is evasive, he says that being married to a restless artist is not as strange as it may seem. His father was a painter, "an old school, macho De Kooning, Pollock kind of character," and his mother a novelist. They were committed New York bohemians, determined to paint and write rather than hold down 9-to-5 jobs. Which explains in large part why in 1968 they moved into the graffiti-covered Masaryk Towers housing project on Columbia Street, just south of Avenue D in Manhattan—it was cheap.

Conley and Jeremijenko met in 1996 at a technological art show in a San Francisco gallery called Blasthaus. At the time, he was doing postdoctorate studies at U.C. Berkeley, and she was at Xerox Parc. A friend had taken him to the gallery, and while admiring an electronic orb that shot off lightning bolts, Conley was astonished when a blond woman strolled by and touched them. She explained the air resistance reduced the voltage to safe levels. "Natalie likes to say there were literally sparks flying when we met," he says.

Speaking of his parents, Conley says, he remembers when he was dating Jeremijenko and his father first met her. She was giving a lecture at the Museum of Modern Art. "I was in California and he called me and said, 'She's a genius.' I felt warm and proud of that. Later, he said we're a perfect couple because she's out of control and I'm a control freak." As for his mother, "I'm scared to admit this, but Natalie and her are very similar. My mom had all sorts of crazy notions. She wanted to start a pet disposal business in New York because she worried that low-income people didn't know what to do with their dogs and cats when they died." It could have been an idea right out of Natalie's head, he says.

Conley admits that he takes as much responsibility as Jeremijenko for the kids' kaleidoscopic, multicultural names—E Harper Nora for their daughter, now eight, and Yo Xing Heyno Augustus Eisner Alexander Weiser Knuckles, now six. They go by "E" and "Yo." Conley explains the names are "interactive" because E can tell everybody her first initial stands for whatever they want it to and because Yo, when he got older, was able to add his own name to the list. He chose "Knuckles," the name of Conley's childhood dog.

Given that Jeremijenko, like matter itself, is constantly in motion, I wonder how she manages as a mother. "It's a big tension in her life," Conley says. "The way she resolves it is to try and integrate family into her work. She's given lectures with a Baby Bjorn on, and sometimes even breastfeeding, which is very funny. As the kids get older, she thinks they'll fit her vision of life as a fluid integration of family and work. She has this ideological idea that there's too much age segregation in our society, and children should be more integrated into adult lives. For me, it's a little too idealistic. A lot of her visions are utopian, and I'm more of a pessimist or a realist. But any two professional folks who don't have a live-in nanny are going to go a little crazy. There's a lot of stress involved. But it's a team. She leans on me pretty heavily when she has an exhibition. We all understand that we won't be seeing much of mommy."

Does he like her work? "I do like her work," he says. "It's completely different than anything I would ever do. Her mind doesn't function the way mine does. I like that. It's just a completely different way of looking at the world. That's the best thing about being together with her. It's always engaging in a way that I wouldn't get anywhere else. She always says things that are surprising to me, even after 10 years. But the price of that is having to live with a mad scientist."

At the same time, Conley admits, because they see the world differently, their work strains their relationship in more ways than one. He means his work too. Last December, Conley wrote what became a controversial Op-Ed (including with *Salon* readers) in the *New York Times*. In the op-ed, titled "A Man's Right to Choose," he declared that if "a father is willing to legally commit to raising a child with no help from the mother he should be able to obtain an injunction against the abortion of the fetus he helped create."

"Natalie was seething mad at me for a month," he says. "I gave it to her to read, but she was too busy. That's the price she had to pay. In every fight we've had since then, I just wait for the line to come up, 'You would chain me to a bed for nine months!' She was so pissed. 'What kind of person am I who would compromise women's autonomy?' She was also personally embarrassed because she thinks it hurt her political credibility in her left-leaning, insular world of science studies and techno-art. She had to deal with the same fallout I did in my academic world and was angry because she was collateral damage."

The marital harmony of the writer and artist, I say, does seems a rather challenging one. Conley nods. The *Times* Op-Ed, he admits, "was largely a response to how much responsibility I've had with the kids. I thought I would be the kind of disengaged, workaholic, distant dad, and

Natalie outdid me on that front, not the disengaged part, but the workaholic part. I also realize it's an incredible gift she gave me. I'm very close to the kids. I arrange the play dates and am deeply engaged in everything they do in a way I might not have been if I were to leave it all to somebody else. Because I'm living with a mad scientist, I often have to be the boring disciplinarian one, and she gets to be the fun, crazy one. It's largely a reversal of gender roles. Sometimes I resent it deeply, and sometimes I really appreciate it."

Not long after, I get a vision of the orchestrated chaos of their life. Their West Village apartment is both cozy home and crowded workplace, books and journals strewn about tables. Dissected parts of Jeremijenko's robotic dogs lie on the kitchen and living room floors. Parakeets and zebra finches perch on a tree in the living room and fly around the apartment. Yo lies on the floor and coils a stuffed toy snake around a stuffed tiger. He might be imitating the animal behavior he once saw on a nature show, except Jeremijenko banned TV from the house four years ago, mostly because, Conley says, she herself got addicted to it, watching cartoons and movies with the kids for hours and crying over the slightest thing. "Natalie weeps at everything," Conley says. "She'll be watching a video and screaming at the screen, 'Swim, Nemo!' I'm like, 'Are you serious?'"

Jeremijenko and Conley have invited friends over for dinner and are trying to find the kids' sweaters and jackets, gather up the food—Conley ordered Pakistan takeout from a place he always sees taxi drivers go to—and move everybody to Jeremijenko's studio two blocks away, where there's more room. After 30 minutes or so of particle-colliding energy, I understand what Conley means when he says that Jeremijenko somehow manages, with two kids, to re-create her experience of growing up in a family of 10 kids. Jeremijenko gathers Yo in her arms, E helps carry some of the bottles of wine, and everybody piles into the street and the New York night.

The intellectual and emotional storm that is Jeremijenko's life may not always be apparent in her art. But it's there. When you think about her cloned trees and talking birds and Hudson River fish buoys, and the extraordinary effort it takes to bring them to the public, you have to marvel at the artist behind them. Working in the nucleus of our increasingly scientific times, she is succeeding, with fantastic irony, to illuminate our humanity, to expose the chain of consequences of our actions and, worse to her, inactions. She's doing it solely by the force of her personality, which, with a few keystrokes on the computer here and turns of a screwdriver there, may be her greatest work. Standing by the Hudson River one afternoon, talking about how the fish will swim through the glowing buoys, she says, "It's a work in progress. That's always part of my story."

Jeff Howe

The Rise of Crowdsourcing

Remember outsourcing? Sending jobs to India and
China is so 2003. The new pool of cheap labor:
everyday people using their spare cycles to create
content, solve problems, even do corporate R&D.

THE PROFESSIONAL

Claudia Menashe needed pictures of sick people. A project
director at the National Health Museum in Washington,
DC, Menashe was putting together a series of interactive
kiosks devoted to potential pandemics like the avian flu. An
exhibition designer had created a plan for the kiosk itself,
but now Menashe was looking for images to accompany the
text. Rather than hire a photographer to take shots of people
suffering from the flu, Menashe decided to use preexisting
images—stock photography, as it's known in the publishing
industry.

In October 2004, she ran across a stock photo collection
by Mark Harmel, a freelance photographer living in Manhattan Beach, California. Harmel, whose wife is a doctor,
specializes in images related to the health care industry.
"Claudia wanted people sneezing, getting immunized, that

sort of thing," recalls Harmel, a slight, soft-spoken 52-year-old.

The National Health Museum has grand plans to occupy a spot on the National Mall in Washington by 2012, but for now it's a fledgling institution with little money. "They were on a tight budget, so I charged them my non-profit rate," says Harmel, who works out of a cozy but crowded office in the back of the house he shares with his wife and stepson. He offered the museum a generous discount: $100 to $150 per photograph. "That's about half of what a corporate client would pay," he says. Menashe was interested in about four shots, so for Harmel, this could be a sale worth $600.

After several weeks of back-and-forth, Menashe emailed Harmel to say that, regrettably, the deal was off. "I discovered a stock photo site called iStockphoto," she wrote, "which has images at very affordable prices." That was an understatement. The same day, Menashe licensed 56 pictures through iStockphoto—for about $1 each.

iStockphoto, which grew out of a free image-sharing exchange used by a group of graphic designers, had undercut Harmel by more than 99 percent. How? By creating a marketplace for the work of amateur photographers—homemakers, students, engineers, dancers. There are now about 22,000 contributors to the site, which charges between $1 and $5 per basic image. (Very large, high-resolution pictures can cost up to $40.) Unlike professionals, iStockers don't need to clear $130,000 a year from their photos just to break even; an extra $130 does just fine. "I negotiate my rate all the time," Harmel says. "But how can I compete with a dollar?"

He can't, of course. For Harmel, the harsh economics lesson was clear: The product Harmel offers is no longer scarce. Professional-grade cameras now cost less than

$1,000. With a computer and a copy of Photoshop, even entry-level enthusiasts can create photographs rivaling those by professionals like Harmel. Add the Internet and powerful search technology, and sharing these images with the world becomes simple.

At first, the stock industry aligned itself against iStockphoto and other so-called microstock agencies like Shutter-Stock and Dreamstime. Then, in February, Getty Images, the largest agency by far with more than 30 percent of the global market, purchased iStockphoto for $50 million. "If someone's going to cannibalize your business, better it be one of your other businesses," says Getty CEO Jonathan Klein. iStockphoto's revenue is growing by about 14 percent a month, and the service is on track to license about 10 million images in 2006—several times what Getty's more expensive stock agencies will sell. iStockphoto's clients now include bulk photo purchasers like IBM and United Way, as well as the small design firms once forced to go to big stock houses. "I was using Corbis and Getty, and the image fees came out of my design fees, which kept my margin low," notes one UK designer in an e-mail to the company. "iStockphoto's micro-payment system has allowed me to increase my profit margin." Welcome to the age of the crowd. Just as distributed computing projects like UC Berkeley's SETI@home have tapped the unused processing power of millions of individual computers, so distributed labor networks are using the Internet to exploit the spare processing power of millions of human brains. The open source software movement proved that a network of passionate, geeky volunteers could write code just as well as the highly paid developers at Microsoft or Sun Microsystems. Wikipedia showed that the model could be used to create a sprawling and surprisingly comprehensive online encyclopedia. And companies like eBay and MySpace have built profitable businesses that couldn't exist without the contributions of users.

All these companies grew up in the Internet age and were designed to take advantage of the networked world. But now the productive potential of millions of plugged-in enthusiasts is attracting the attention of old-line businesses, too. For the last decade or so, companies have been looking overseas, to India or China, for cheap labor. But now it doesn't matter where the laborers are—they might be down the block, they might be in Indonesia—as long as they are connected to the network.

Technological advances in everything from product design software to digital video cameras are breaking down the cost barriers that once separated amateurs from professionals. Hobbyists, part-timers, and dabblers suddenly have a market for their efforts, as smart companies in industries as disparate as pharmaceuticals and television discover ways to tap the latent talent of the crowd. The labor isn't always free, but it costs a lot less than paying traditional employees. It's not outsourcing; it's crowdsourcing.

It took a while for Harmel to recognize what was happening. "When the National Health Museum called, I'd never heard of iStockphoto," he says. "But now, I see it as the first hole in the dike." In 2000, Harmel made roughly $69,000 from a portfolio of 100 stock photographs, a tidy addition to what he earned from commissioned work. Last year his stock business generated less money—$59,000— from more than a thousand photos. That's quite a bit more work for less money.

Harmel isn't the only photographer feeling the pinch. Last summer, there was a flurry of complaints on the *Stock Artists Alliance* online forum. "People were noticing a significant decline in returns on their stock portfolios," Harmel says. "I can't point to iStockphoto and say it's the culprit, but it has definitely put downward pressure on prices." As a result, he has decided to shift the focus of his business to assignment work. "I just don't see much of a future for professional stock photography," he says.

THE PACKAGER

"Is that even a real horse? It looks like it doesn't have any legs," says Michael Hirschorn, executive vice president of original programming and production at VH1 and a creator of the cable channel's hit show *Web Junk 20*. The program features the 20 most popular videos making the rounds online in any given week. Hirschorn and the rest of the show's staff are gathered in the artificial twilight of a VH1 editing room, reviewing their final show of the season. The horse in question is named Patches, and it's sitting in the passenger seat of a convertible at a McDonald's drive-through window. The driver orders a cheeseburger for Patches. "Oh, he's definitely real," a producer replies. "We've got footage of him drinking beer." The crew breaks into laughter, and Hirschorn asks why they're not using that footage. "Standards didn't like it," a producer replies. Standards—aka Standards and Practices, the people who decide whether a show violates the bounds of taste and decency—had no such problem with Elvis the Robocat or the footage of a bicycle racer being attacked by spectators and thrown violently from a bridge. *Web Junk 20* brings viewers all that and more, several times a week. In the new, democratic age of entertainment by the masses, for the masses, stupid pet tricks figure prominently.

The show was the first regular program to repackage the Internet's funniest home videos, but it won't be the last. In February, Bravo launched a series called *Outrageous and Contagious: Viral Videos,* and USA Network has a similar effort in the works. The E! series *The Soup* has a segment called "Cybersmack," and NBC has a pilot in development hosted by Carson Daly called *Carson Daly's Cyberhood,* which will attempt to bring beer-drinking farm animals to the much larger audiences of network TV. Al Gore's Current TV is placing the most faith in the model: More than 30 percent of its programming consists of material submitted by viewers.

Viral videos are a perfect fit for VH1, which knows how to repurpose content to make compelling TV on a budget. The channel reinvented itself in 1996 as a purveyor of tawdry nostalgia with *Pop-Up Video* and perfected the form six years later with *I Love the 80s.* "That show was a good model because it got great ratings, and we licensed the clips"—quick hits from such cultural touchstones as *The A-Team* and *Fatal Attraction*—"on the cheap," Hirschorn says. (Full disclosure: I once worked for Hirschorn at Inside.com.) But the C-list celebrity set soon caught on to VH1's searing brand of ridicule. "It started to get more difficult to license the clips," says Hirschorn, who has the manner of a laid-back English professor. "And we're spending more money now to get them, as our ratings have improved."

But Hirschorn knew of a source for even more affordable clips. He had been watching the growth of video on the Internet and figured there had to be a way to build a show around it. "I knew we offered something YouTube couldn't: television," he says. "Everyone wants to be on TV." At about the same time, VH1's parent company, Viacom, purchased iFilm—a popular repository of video clips—for $49 million. Just like that, Hirschorn had access to a massive supply of viral videos. And because iFilm already ranks videos by popularity, the service came with an infrastructure for separating the gold from the god-awful. The model's most winning quality, as Hirschorn readily admits, is that it's "incredibly cheap"—cheaper by far than anything else VH1 produces, which is to say, cheaper than almost anything else on television. A single 30-minute episode costs somewhere in the mid-five figures—about a tenth of what the channel pays to produce *so noTORIous,* a scripted comedy featuring Tori Spelling that premiered in April. And if the model works on a network show like *Carson Daly's Cyberhood,* the savings will be much greater: The average half hour of network TV comedy now costs nearly $1 million to produce.

Web Junk 20 premiered in January, and ratings quickly exceeded even Hirschorn's expectations. In its first season, the show is averaging a respectable half-million viewers in the desirable 18-to-49 age group, which Hirschorn says is up more than 40 percent from the same Friday-night time slot last year. The numbers helped persuade the network to bring *Web Junk 20* back for another season.

Hirschorn thinks the crowd will be a crucial component of TV 2.0. "I can imagine a time when all of our shows will have a user-generated component," he says. The channel recently launched *Air to the Throne,* an online air guitar contest, in which viewers serve as both talent pool and jury. The winners will be featured during the *VH1 Rock Honors* show premiering May 31. Even VH1's anchor program, *Best Week Ever,* is including clips created by viewers.

But can the crowd produce enough content to support an array of shows over many years? It's something Brian Graden, president of entertainment for MTV Music Networks Group, is concerned about. "We decided not to do 52 weeks a year of *Web Junk,* because we don't want to burn the thing," he says. Rather than relying exclusively on the supply of viral clips, Hirschorn has experimented with soliciting viewers to create videos expressly for *Web Junk 20.* Early results have been mixed. Viewers sent in nearly 12,000 videos for the Show Us Your Junk contest. "The response rate was fantastic," says Hirschorn as he and other staffers sit in the editing room. But, he adds, "almost all of them were complete crap."

Choosing the winners, in other words, was not so difficult. "We had about 20 finalists." But Hirschorn remains confident that as user-generated TV matures, the users will become more proficient and the networks better at ferreting out the best of the best. The sheer force of consumer behavior is on his side. Late last year the Pew Internet & American Life Project released a study revealing that 57 percent of 12-to 17-year-olds online—12 million individuals—are creating content of some sort and posting it to the Web. "Even if the signal-to-noise ratio never improves— which I think it will, by the way—that's an awful lot of good material," Hirschorn says. "I'm confident that in the end, individual pieces will fail but the model will succeed."

THE TINKERER

The future of corporate R&D can be found above Kelly's Auto Body on Shanty Bay Road in Barrie, Ontario. This is where Ed Melcarek, 57, keeps his "weekend crash pad," a one-bedroom apartment littered with amplifiers, a guitar, electrical transducers, two desktop computers, a trumpet, half of a pontoon boat, and enough electric gizmos to stock a RadioShack. On most Saturdays, Melcarek comes in, pours himself a St. Remy, lights a Player cigarette, and attacks problems that have stumped some of the best corporate scientists at Fortune 100 companies.

Not everyone in the crowd wants to make silly videos. Some have the kind of scientific talent and expertise that corporate America is now finding a way to tap. In the process, forward-thinking companies are changing the face of R&D. Exit the white lab coats; enter Melcarek—one of over 90,000 "solvers" who make up the network of scientists on InnoCentive, the research world's version of iStockphoto.

Pharmaceutical maker Eli Lilly funded InnoCentive's launch in 2001 as a way to connect with brainpower outside the company—people who could help develop drugs and speed them to market. From the outset, InnoCentive threw open the doors to other firms eager to access the network's trove of ad hoc experts. Companies like Boeing, DuPont, and Procter & Gamble now post their most ornery scientific problems on InnoCentive's Web site; anyone on InnoCentive's network can take a shot at cracking them.

The companies—or seekers, in InnoCentive parlance—pay solvers anywhere from $10,000 to $100,000 per solution. (They also pay InnoCentive a fee to participate.) Jill Panetta, InnoCentive's chief scientific officer, says more than 30 percent of the problems posted on the site have been cracked, "which is 30 percent more than would have been solved using a traditional, in-house approach."

The solvers are not who you might expect. Many are hobbyists working from their proverbial garage, like the University of Dallas undergrad who came up with a chemical to use in art restoration and the Cary, North Carolina, patent lawyer who devised a novel way to mix large batches of chemical compounds.

This shouldn't be surprising, notes Karim Lakhani, a lecturer in technology and innovation at MIT, who has studied InnoCentive. "The strength of a network like InnoCentive's is exactly the diversity of intellectual background," he says. Lakhani and his three coauthors surveyed 166 problems posted to InnoCentive from 26 different firms. "We actually found the odds of a solver's success increased in fields in which they had no formal expertise," Lakhani says. He has put his finger on a central tenet of network theory, what pioneering sociologist Mark Granovetter describes as "the strength of weak ties." The most efficient networks are those that link to the broadest range of information, knowledge, and experience.

Which helps explain how Melcarek solved a problem that stumped the in-house researchers at Colgate-Palmolive. The giant packaged goods company needed a way to inject fluoride powder into a toothpaste tube without it dispersing into the surrounding air. Melcarek knew he had a solution by the time he'd finished reading the challenge: Impart an electric charge to the powder while grounding the tube. The positively charged fluoride particles would be attracted to the tube without any significant dispersion.

"It was really a very simple solution," says Melcarek. Why hadn't Colgate thought of it? "They're probably test tube guys without any training in physics." Melcarek earned $25,000 for his efforts. Paying Colgate-Palmolive's R&D staff to produce the same solution could have cost several times that amount—if they even solved it at all. Melcarek says he was elated to win. "These are rocket-science challenges," he says. "It really reinforced my confidence in what I can do." Melcarek, who favors thick sweaters and a floppy fishing hat, has charted an unconventional course through the sciences. He spent four years earning his master's degree at the world-class particle accelerator in Vancouver, British Columbia, but decided against pursuing a PhD. "I had an offer from the private sector," he says, then pauses. "I really needed the money." A succession of "unsatisfying" engineering jobs followed, none of which fully exploited Melcarek's scientific training or his need to tinker. "I'm not at my best in a 9-to-5 environment," he says. Working sporadically, he has designed products like heating vents and industrial spray-painting robots. Not every quick and curious intellect can land a plum research post at a university or privately funded lab. Some must make HVAC systems.

For Melcarek, InnoCentive has been a ticket out of this scientific backwater. For the past three years, he has logged onto the network's Web site a few times a week to look at new problems, called challenges. They are categorized as either chemistry or biology problems. Melcarek has formal training in neither discipline, but he quickly realized this didn't hinder him when it came to chemistry. "I saw that a lot of the chemistry challenges could be solved using electro-mechanical processes I was familiar with from particle physics," he says. "If I don't know what to do after 30 minutes of brainstorming, I give up." Besides the fluoride injection challenge, Melcarek also successfully came up with a method for purifying silicone-based solvents. That challenge paid $10,000. Other Melcarek solutions have been close runners-up, and he currently has two more up for consideration. "Not bad for a few weeks' work," he says with a chuckle.

It's also not a bad deal for the companies that can turn to the crowd to help curb the rising cost of corporate research. "Everyone I talk to is facing a similar issue in regards to R&D," says Larry Huston, Procter & Gamble's vice president of innovation and knowledge. "Every year research budgets increase at a faster rate than sales. The current R&D model is broken."

Huston has presided over a remarkable about-face at P&G, a company whose corporate culture was once so insular it became known as "the Kremlin on the Ohio." By 2000, the company's research costs were climbing, while sales remained flat. The stock price fell by more than half, and Huston led an effort to reinvent the way the company came up with new products. Rather than cut P&G's sizable in-house R&D department (which currently employs 9,000 people), he decided to change the way they worked.

Seeing that the company's most successful products were a result of collaboration between different divisions, Huston figured that even more cross-pollination would be a good thing. Meanwhile, P&G had set a goal of increasing the number of innovations acquired from outside its walls from 15 percent to 50 percent. Six years later, critical components of more than 35 percent of the company's initiatives were generated outside P&G. As a result, Huston says, R&D productivity is up 60 percent, and the stock has returned to five-year highs. "It has changed how we define the organization," he says. "We have 9,000 people on our R&D staff and up to 1.5 million researchers working through our external networks. The line between the two is hard to draw."

P&G is one of InnoCentive's earliest and best customers, but the company works with other crowdsourcing networks as well. YourEncore, for example, allows companies to find and hire retired scientists for one-off assignments. Nine-Sigma is an online marketplace for innovations, matching seeker companies with solvers in a marketplace similar to InnoCentive. "People mistake this for outsourcing, which it most definitely is not," Huston says. "Outsourcing is when I hire someone to perform a service and they do it and that's the end of the relationship. That's not much different from the way employment has worked throughout the ages. We're talking about bringing people in from outside and involving them in this broadly creative, collaborative process. That's a whole new paradigm."

THE MASSES

In the late 1760s, a Hungarian nobleman named Wolfgang von Kempelen built the first machine capable of beating a human at chess. Called the Turk, von Kempelen's automaton consisted of a small wooden cabinet, a chessboard, and the torso of a turbaned mannequin. The Turk toured Europe to great acclaim, even besting such luminaries as Benjamin Franklin and Napoleon. It was, of course, a hoax. The cabinet hid a flesh-and-blood chess master. The Turk was a fancy-looking piece of technology that was really powered by human intelligence. Which explains why Amazon.com has named its new crowdsourcing engine after von Kempelen's contraption. Amazon Mechanical Turk is a Web-based marketplace that helps companies find people to perform tasks computers are generally lousy at—identifying items in a photograph, skimming real estate documents to find identifying information, writing short product descriptions, transcribing podcasts. Amazon calls the tasks human intelligence tasks (HITs); they're designed to require very little time, and consequently they offer very little compensation—most from a few cents to a few dollars.

InnoCentive and iStockphoto are labor markets for specialized talents, but just about anyone possessing basic literacy can find something to do on Mechanical Turk. It's crowdsourcing for the masses. So far, the program has a mixed track record: After an initial burst of activity, the amount of work available from requesters—companies offering work on the site—has dropped significantly. "It's gotten a little gimpy," says Alan Hatcher, founder of Turker Nation, a community forum. "No one's come up with the killer app yet." And not all of the Turkers are human: Some would-be workers use software as a shortcut to complete the tasks, but the quality suffers. "I think half of the people signed up are trying to pull a scam," says one requester who asked not to be identified. "There really needs to be a way to kick people off the island."

Peter Cohen, the program's director, acknowledges that

Mechanical Turk, launched in beta in November, is a work in progress. (Amazon refuses to give a date for its official launch.) "This is a very new idea, and it's going to take some time for people to wrap their heads around it," Cohen says. "We're at the tippy-top of the iceberg."

A few companies, however, are already taking full advantage of the Turkers. Sunny Gupta runs a software company called iConclude just outside Seattle. The firm creates programs that streamline tech support tasks for large companies, like Alaska Airlines. The basic unit of iConclude's product is the repair flow, a set of steps a tech support worker should take to resolve a problem.

Most problems that iConclude's software addresses aren't complicated or time consuming, Gupta explains. But only people with experience in Java and Microsoft systems have the knowledge required to write these repair flows. Finding and hiring them is a big and expensive challenge. "We had been outsourcing the writing of our repair flows to a firm in Boise, Idaho," he says from a small office overlooking a Tully's Coffee. "We were paying $2,000 for each one."

As soon as Gupta heard about Mechanical Turk, he suspected he could use it to find people with the sort of tech support background he needed. After a couple of test runs, iConclude was able to identify about 80 qualified Turkers, all of whom were eager to work on iConclude's HITs. "Two of them had quit their jobs to raise their kids," Gupta says. "They might have been making six figures in their previous lives, but now they were happy just to put their skills to some use."

Gupta turns his laptop around to show me a flowchart on his screen. "This is what we were paying $2,000 for. But this one," he says, "was authored by one of our Turkers." I ask how much he paid. His answer: "$5."

Emily Nussbaum

Mothers Anonymous

In the collective id known as UrbanBaby, New
York women confess their darkest fears about
parenting and marriage—and, not infrequently,
go to war over them.

11:37 a.m.

"I just found out that my DH [dear husband] is
cheating on me while he's away in europe. I have
an email from the woman planning additional time
together. I don't want to continue a life with a
cheater . . . feel so sick and lonely. what do i
do now?"

10:24 p.m.

"I drink. I love it. It is my best friend
sometimes. but other times it is my enemy. I get
so lonely sometimes and it quells me. anybody here
experience the same thing?"

8:21 p.m.

"I'm a sahm with 2 dc [dear children]. I have a
cushy life with a housekeeper once a week and big
budget for stuff. Still I'm so bored and lonely
with life. I eat and go on the computer all the
time. My dh works all the time too."

8:22 p.m.

"Grow up."

8:26 p.m.

"I'm 40, how much can I grow up at this point?"

A SHORT HISTORY OF AMERICAN MOTHERHOOD

Once upon a time, becoming a mother was something you did alone, in your home, with your baby. Your sources of expertise were few: women in your family, women on your block, and your doctor. Your husband knew nothing, and when you had a question, there was Dr. Spock. Maybe you were happy, or angry, or drunk, or overwhelmed, or pleasantly bored, or deeply satisfied. But those emotions lived at home.

Then came the Internet. (Okay, then came feminism. But after that, the Internet.) And for New York mothers, then came UrbanBaby.

If you're a mother of a certain type—upscale and analytical—you have likely heard of the site. It's a discussion board, but it's also a bit of an obsession and a bit of a drain, in the sense that it's a place where a lot of New York mothers dump their most toxic feelings, in maroon-and-bright-green threads that spill down the screen with little organization. On UrbanBaby, the private lives of city mothers are lit up and exposed. All the houses are glass there, and everybody's got a rock.

In part, this is because UrbanBaby is anonymous—and online, anonymity acts like a combination of a truth serum and a very strong cocktail. But this is also because being a mother can feel like sitting in a solemn lecture room, listening and taking notes, and repressing impulse after impulse to yell out dirty words. On UrbanBaby, people blurt out these dirty words.

On my screen right now, these threads co-exist uneasily:

"What percentage of people do you think just 'settle', i.e. just marry the person they are with when they get to a certain age not b/c they think they are the perfect person for them but are too scared to go out there again?" "If you saw a jewish girl wearing a vineyard vines sweater what would you think? be 100% honest."
"Why do I have to keep telling dh size doesn't matter? why?????!!!! I hate lying."
"Is Anderson Cooper gay?"
"Post names of really crappy nannies you know."
"These soycrisps are addictive."
"I am in love with my daughter. She is on this new roar and growl routine as she crawls like a crab across the floor. I could watch this forever."
"Tell us a secret about yourself. Something you don't want other people to know. We won't tell a soul."

MANHATTAN MOTHER

Everyone knows the scary archetype of the monster Manhattan mother: She's all elbows and no bosom, like ritzy Mrs. X in *The Nanny Diaries* or careerist Kate Hudson in *Raising Helen;* she's every East Side matron on *Wife Swap* braying about "me time." Which is to say, a professional who treats her child like a résumé; a fashionista who wears her child as an accessory; a trophy wife who leverages her child as an excuse to quit work and go shopping. Or perhaps she's turned motherhood itself into her career, driving her child insane with flash cards, like the Parker Posey character in *Best of Show,* except with a toddler instead of a purebred Weimaraner: obsessed with getting exactly the right plush toy, right now.

Even many actual New York moms talk about one another this way, so much have we internalized the notion that there is something loathsome and prissy and spoiled at heart about New York motherhood. Something at once neglectful and overprotective. Something not very motherly at all.

But then again, I'm one of them. From the time I was three months pregnant, I was online, researching. I bookmarked sites for music classes and cribs. I lurked around, the way people do, surfing from place to place. But something about UrbanBaby kept drawing me back. The tone was different. It was not supportive—a welcome shock. It was toxic but also compassionate in surprising moments. It was an antidote to sites like BabyCenter—those earnest malls trafficking in humor-free LOLs and "babydust" (a virtual gift sent to women who are TTC, or "trying to conceive.")

You'd think the board's randomness would make reading it unbearable. And maybe it is. But for many, including me, it is also irresistible, a place for New York mothers to talk about the most taboo subjects—money, sex, and especially marriage, the ones rotting from the inside, the ones where the couples never talk and never have sex, the ones organized along occult power lines invisible to the outside world. You don't have to be a very voyeuristic person to be drawn in, but it helps. On UrbanBaby, women openly confess to an ambivalence about parenting that no one is allowed to admit in the sunlight. They disclose financial secrets they'd never discuss with their closest friends and secrets about their feelings toward their husbands, which sometimes amounts to the same thing.

They blurt out entire novels about their lives in staccato dialogues.

7:40 p.m.
"I never wanted kids and now I am pregnant."
7:42 p.m.
"do you want this one?"
7:44 p.m.
"'Want' is such a strong word. My husband wants
it. Sometimes I think it's interesting. I am
EXTREMELY uncomfortable (2 months to go) and I am
BORED."

UrbanBaby began as a New York project, founded by former *Esquire* editor Susan Maloney and her husband, John, for women who "couldn't relate to pink and blue," and though it has since gone national, its flavor is still very New York. There are lists of lactation consultants and interviews with "Moms About Town" like Nicole Miller; there are aspirational illustrations of skinny moms with posh diaper bags. A Craigslist-ish area allows moms to hawk gently used Bugaboos.

But as with so many online phenomena, what really took off about the site wasn't the useful part. It was the message board, which flourished not despite but because of its chaotic design. On UrbanBaby, no one has a profile or a screen name, so there's no telling who is posting from thread to thread—could be Mary Louise Parker, could be your bitchy neighbor. Comments are studded with baffling abbreviations like dh (dear husband), dd (dear daughter), ds (dear son), and BTDT (been there done that) and jargon like "sanctimommy" (a self-righteous mom) and "*Über-boober*" (a self-righteous mom obsessed with breast-feeding).

Amid all this, invisible "deleters" erase entire threads for no clear reason. (Were they too sexual? Too political? Did they talk about the deleters too much, à la *Fight Club*?)

And if you are banned, you will get nothing but a mysterious message telling you how long you need to wait before being allowed to post. In seconds. Like this: "You will not be able to post for 24,083 seconds." No explanation.

As women post en masse over the course of the day and long into the night, the mood changes: The daylight crowd tends to be prissier; the night crowd rowdier (and drunker); the late-night crowd surrealistic and unpredictable, made up of the extremely sleep deprived, from mothers of newborns to insomniacs in the midst of a divorce. But certain shared obsessions loop back, sometimes for years on end. There's the conflict between Park Slope (crunchy, sanctimonious) and the Upper East Side (elitist, spoiled). There are "Muffintops," a.k.a. the postpartum love handles, and "Frenemies," or backbiting false friends. Celebrity moms— Gwyneth, Brit, SJP, MLP—and their failings and whether they are secretly on UrbanBaby. Stay-at-home moms: How much help do they have? Cry it out: Is it cruel to sleep-train your child? Circumcision, extended breast-feeding, autism, PPD (postpartum depression), BPPs (bitter poor persons), and TT (top-tier) preschools.

There are the notorious "bad nanny sightings":

"If you know the parents of this beautiful girl please tell them that their nanny is verbally abusive to the baby and let her scream today for at least 45 minutes and kept telling the baby 'you have to learn.'"

There are recurrent escape plans:

"I have a secret fantasy of splitting with dh and taking db [dear baby] and leaving him with dd. He's making her such a spoiled brat, I'm tired of being the bad guy, and I'm tired of his shit." ·

And there are endless comparative polls: How much do you weigh? What's your HHI (household income)? Your ring size, your bra size, your thread weave? Someone calling herself Psychic Mom tries to predict the future, and Hong Kong Mom posts in the wee hours of the morning to sleepless mothers in the East Village. A tough-love Marxist urges a mom whose nanny has quit without notice to "walk it off." Muslim moms debate identity politics with Jewish moms. A self-proclaimed Life Coach mom offers to solve everyone's problems at once.

Once you're hooked, it's very hard to log off. So hard that women have been known to get banned deliberately just so they can break their addiction. And there are some very mean women out there—women who will accuse a woman whose child has died of making the whole thing up, women who will attack another woman's child's name until she posts back that she's crying. The site is poisonous at times but also strangely comforting and frequently hilarious: There are other women out there who are all wound up; cracking bizarre running jokes; and overthinking everything, but overthinking it together.

And if you ever wanted to know what people were saying behind your back, here is your answer.

DID YOU MARRY FOR MONEY?

And then there is marriage, the true and hidden subject of much of the site, which is so often, and startlingly, less about babies themselves than about what babies have done to these women's relationships. Numerous posts concern cheating husbands—wives snooping via e-mail; wives finding out that their husbands have online-dating accounts; wives getting "key-logger" programs to trace the patterns of his computer use; the recurrent, fiery debate about whether investment bankers cheat more than other husbands, whether porn and strippers constitute or indicate cheating, whether a husband is cheating with his besotted single-woman colleague who keeps sending him text messages, and whether anyone knows a good divorce lawyer.

See also "Have you ever had an affair?"

See also "Did you marry for money?"

There's a melancholic subtheme about nostalgia for ex-boyfriends, especially erotic dreams about ex-boyfriends during pregnancy. A cadre of "EWGs"—ex–working girls—inform the horrified uptown moms what their men are really like. There are a lot of questions about gayness (Is my husband gay? Is my child gay?) and a recurrent debate about whether enjoying *Brokeback Mountain* is a sign of gayness. A lot of the time, the women simply post on and on about what their husbands don't do: They don't help out, they won't be kind, they never come home until the wee hours, and they demand sex the mothers don't want to give and refuse sex the mothers beg for.

On Mother's Day, a day I had expected to be a happy one on UrbanBaby, the board scrolled by all day with disappointment: no *card,* no *gift,* no *help,* a weekend spent with a terrible mother-in-law and a husband who claimed Mother's Day wasn't Wives' Day, so he didn't need to get his wife anything.

And amid all the self-pity, there's a loathing of self-pity. "A baby is tiring, just suck it up or don't have kids," replied one poster to a mother of a three-week-old who wondered if other women's husbands helped with night feedings.

PUSH PRESENTS

There are phenomena I'd never heard of before I read UrbanBaby. For instance, "push presents." Apparently, it's a tradition to give a fancy piece of jewelry—a diamond tennis bracelet, say, or an expensive ring—to your wife as a reward, or perhaps a motivation, for getting through labor. On UrbanBaby, it's also a tradition to mock these bracelets. And to compare them.

When bonus time comes around for the wives of IBers (investment bankers) and "BigLaw" lawyers, the boards go mad with Schadenfreude and envy and rage. In one such discussion, a woman explained that she couldn't stay home even if she wanted to because she needed her salary: She made $150,000 a year. "Not significant," responded the other poster. Other people began to chime in: "150,000 per year is not significant income? What planet do you live on?" And "I would quit in a heartbeat if I made $150K. It isn't significant to me."

On the boards of UrbanBaby, the economic calculus of Manhattan and Brooklyn (and sometimes Queens and often Long Island) is hashed out with a cruel candor that is nearly impossible to find in other places, if only because women in disparate economic circumstances are forced to confront one another's experiences head-on. Corporate bigwigs post there; so do their nannies. Single mothers sinking into debt hash out their budgets in public; so do women in marriages where both parties float on a sea of family money and never work and spend their time managing their investments. Perhaps unsurprisingly, the sharpest clashes happen not between poor women and rich women but those separated by the slimmest difference—the anxious perforation between the wealthy and the super-wealthy.

"DH and I make about 300K per year. We have 200K in savings. Can we afford a $1 million apartment? We can't believe we have to spend that much to get anything close to what we want, and we both grew up fairly blue collar, so it just seems like an absurd amount of money for us (despite decent income). Have I just not adjusted to the way things are now?"

Maybe it's no wonder that on UrbanBaby, emotions are so close to the surface. With a 26 percent rise in the number of 5-year-olds in New York from 2000 to 2004, a simultaneous rise in the percentage of families staying in the city instead of moving to the suburbs, and an increase in the number of mothers choosing to stay home with their children, the island seems to be filling up with strollers at precisely the moment when the sidewalks have narrowed. Who can afford to have one child, let alone two or three or four? It's little wonder that motherhood on UrbanBaby is surprisingly hard to distinguish from class war.

Of particular fascination to the UrbanBaby women are those mothers whose economic status seems to many posters both enviable and ridiculous: stay-at-home moms with full-time nannies and expensive tastes.

11:07 a.m.
"my wohm [work-outside-the-home mom] friend is always muttering sarcastically how it must be nice as a sahm [stay-at-home mom] to have lunch with friends & hang out at the park. So I told her it must be nice to afford designer clothes & go on luxurious vacations 2x a yr. Now she is mad at me! WTF?"
11:08 a.m.
"both of you steer clear of me, I'm SAHM who lunches and I have designer clothes and take luxe vacations."
11:08 a.m.
"la di da."

One night, a woman posts this seemingly nonrhetorical question: "If your dh had a 5mil trust fund would you stay home? 2 kids and dh does not work." Responses range from a deadpan "uh, yeah" to "someone has to work . . . 5 mil is not enough for forever." A long thread branching off examines the premise that a trust fund providing interest of $350,000 to $500,000 is not enough to live on. "Not enough for whom?" asked one poster incredulously. Another poster replies, "Me. We currently live a 15k/month lifestyle, net, with 1 dc and no school costs"—and then promptly summarizes her expenses for an invisible audience: "7k rent, 1k PT sitter, eating out 1.5–2k, utilities 500, travelling 2k, clothing 1k, out and about 'cash' 1k."

Yet although there's a lot of talk about money—including inane semiotic detanglings of the differences between old and new money—just as often the financial nitpicking is submerged into chitchat about other differences, from breast-feeding to baby nurses. And so the women focus on the five pounds that separate them from the woman who shares their bench in the playground. And everything, from a serving of YoBaby yogurt to the precise numerical factor that mothers should spend playing with their children (Thirty minutes a day? Thirty minutes an hour?), is pulled into the calculus of how good a mother you are. Strangely, the sensation one gets from the most privileged mothers on the boards is of constant scarcity. Only so many children can get into a "top-tier school," someone else's toddler has more words than yours, other people are taking vacations so vastly superior to your vacations that your vacations barely count as vacations.

And maybe people go to UrbanBaby to be judged in the first place. Certainly, there is something masochistic about the experience, because no matter what you do, someone will disapprove, starting with the day you give birth. Refuse the epidural, and someone will sneer that you're a hippie fool—beg for the drugs, and someone else will suggest you're a weakling. Home-birth? Lunatic. Scheduled C-section? Control freak. IVF? Unnatural. You can hire a doula and a midwife and a lactation consultant and be called a flake, or hire a nanny, a night nurse, and a maid and be called a spoiled brat. Or you can hire no one and get called a "martyr mom."

Perversely, this is a comfort: If there are no right choices, there are no wrong ones either.

SEARCHING FOR "BITTERNESS"

Which is not to say that there isn't supportiveness on Urban-Baby. On the boards devoted to pregnancy and newborns and women trying to conceive, you can find answers to every question, immediately, from a chorus of helpful invisible fellow mothers, at any hour. Much of the advice is witty and precise: People will recommend toys, help you choose an OB/GYN, reassure you after a miscarriage. Last Monday, a mother posted, "I feel like I was a bully this am with dc. I was yelling, dd was in tears, ds (4) said 'You made me upset and I hope you're happier when you pick me up,'" and her invisible friends commiserated in a flash: "that was me yesterday" (two minutes later), and "i hate when that happens. i went through a couple of weeks like that & finally told myself just to snap out of it" (four minutes later.) On UrbanBaby, nothing is too ordinary to discuss, and nothing is too abstract—and it all just keeps flowing by, swerves in the river of the shared mother mind.

Mothers post about happiness as well, often defensively invoking another parenting site: the cheery, cloying Baby-Center. "I love my muffin so much it hurts. New mother here to 6 mo ds and I still can't believe this amazing little thing is my son. Ok, sorry for the babycenter moment. Carry on!"

But sometimes, the dark side of UrbanBaby is more alluring. It's what keeps me lurking on the craziest, most surrealistic boards—especially the toddler board, which attracts the most outrageous, least child-centered posting. It's there that you can feel like you have tapped into that cartoon fantasy of the Manhattan Mother, or at least her startling real-life analogue, a woman who may not always be likable but who is full of complicated conflicts and contradictions.

Some nights when my baby is sleeping, instead of reading the boards straight through, I play with the site's primitive search function to see what turns up. One night, I searched for *bitter* and found this stay-at-home mom posting in a rage: "I am so angry and bitter. There is no solution." Her baby woke up at 5:22, her husband won't wake up for night feedings, and "he gets to do whatever he wants/whenever he wants. Me? Never and going back to work wouldn't change a thing. It would be me rushing home to relieve nanny. and it would still be *me* on weekend. I truly don't foresee marriage surviving children . . . The rage I feel is unreal. Yes, he makes a lot of money and allows me to SAH. but that is about it."

Then I tried *cheating.*

"Can anybody help me think about this—I think I have to leave my husband (he's been cheating on me and worse) but i feel sad about having to date people that don't love my daughter in the same way he does. How does one do that?"

Other good search words: *pathetic, furious, divorce, whore, affair,* and *nanny.*

But really, anything can work. Try *behavior:*

"Please help," writes one poster. "I know that dh is cheating on me, but I love the person that he's become. He sings around the house, loves playing with the kids, and is a general joy to be around. It's just so sad that I couldn't be the woman to inspire this new behavior."

On UrbanBaby, two equal and opposite forms of nostalgia nudge up against one another like hot and cold fronts.

First, there's nostalgia for the 1950s, when (the fantasy goes) mothers were a safe, proscribed, protected, narrow, and paid-for class. There weren't many choices for women, and that was a good thing. And at the center of this fantasy is a man who is a good provider, someone who can bring the 1950s life into the far-less-affordable 2006.

Numerous posts play out this fantasy like that old dating board game, in which the fate of your plastic marker was determined by which man you landed on. "Would you rather marry a blue collar guy who makes a lot of $$, a white collar who makes little money, or old money who is a cheater?" asks one post. "If you were to marry again, which would you prefer: doctor, lawyer, ib'er, hf'er [hedge funder], journalist?" asks another. (This provokes a certain amount of hilarity at the limited choices available.)

In some posts, there's a giddy, anxious glee about reclaiming those roles. "I am one sick little 50's housewife. *Love* making a menu, list of ingredients, clipping coupons, going to store, buying specials, stocking up, coming in under budget. *Love it!* What the hell is wrong with me??" And sometimes there's something else: a feeling of loneliness, a rush of rage at not being appreciated. Stay-at-home moms on the site often seem amazingly angry, and it's not unusual to read a post that says nothing but "I hate my husband."

And then there's nostalgia for the 1970s, when (the fantasy goes) motherhood was a freewheeling, chain-smoking, martini-swilling, no-car-seat experience, and we all came out just *fine.*

```
2:15 p.m.
"I remember sliding around on the vinyl backseat
of my grandparents' dodge swinger."
2:17 p.m.
"Fun, right? I remember getting a dozen kids in
the back of a neighbor's pickup to go swimming at
the unattended lake."
```

In one post, a woman suggests that people imagine posts that would have been on UrbanBaby if it had existed in 1970 and got an array of parody postings from "My dumb kid ate all my dope!" to "My wooden spoon broke as i was spanking dc—should I just use my hand to spank?" to "My 3 yo twins ride their Big Wheels and Hippity Hops across the street to the park alone, should I make them come home before dark?"

BRITNEY

The night Britney Spears spoke to Matt Lauer to defend herself as a mother, the boards were on fire. A hundred (or maybe 20, or maybe 500: It's impossible to tell on Urban-Baby) invisible women live-blogged their response to her every word, critiquing the eyelash that dangled from her face like a broken windshield wiper; laughing at her description of her husband, Kevin Federline, as "simple"; moaning in disbelief when she defended her driving her child around without a car seat as "country."

She was everything that they (or we) were not. She was tacky. She was "white trash." She gave her child a "Wal-Mart" name. She definitely didn't seem to be thinking too much.

And yet people also felt sorry for her. They were angry

at her—she was a bad mother!—but they could also identify. "Does anyone ever wonder if celebs post here (like Britney for example) and then feel bad about what people say
here about her. I would be so horrified if there were somewhere where people criticized my parenting!"

And "I have been thinking for a long time that even
though Britney does some stupid things it's sweet how much
she holds Tater Tot."

Within days, people began admitting to their inner Britney:

*"Call me Britney. I was carrying dd at the park
this weekend, slipped on some water & totally
wiped out . . . I conked my head & everything &
now I can't stop having these horrible images of
what could have happened."*

*"I think I totally get why britney stuffed herself
in that 'cute figure' outfit last night: she
hasn't grasped yet that she isn't still the super-
hot-bodied girl anymore. I only know from my own
personal experience—not that I wore jean skirts,
but it has taken me until after my second baby was
born to get it through my head that my body was
not as tight as it used to be. it's like I was
seeing the old me in the mirror out of habit."*

Celebrities are huge on UrbanBaby—the boards often
clutter up with posts reading "TEAM JOLIE!" and
"TEAM ANISTON!"—but of all the famous women
whose alternately stuffed and unstuffed wombs have been
documented in *US Weekly,* Britney brings out the strongest
emotions. She's the much-mocked public mother, the one
everyone was afraid lurked inside them, the woman with
the bull's-eye on her belly. She was weeping and a mess and
totally out of control, and that was the scariest thing of all.

Late at night, the women on UrbanBaby start talking dirty. They list how many partners they have had. They argue about oral sex and start up extended chats about lesbian seductions—including confessions of playground affairs that seem at once convincing (details about children napping in a Pack 'n Play!) and suspiciously akin to Penthouse's "Forum" stories. The next morning, the discussion has been deleted.

And then there are their opposite numbers. Search "sexless" and you find a series of posts seeking community: "Anyone else in a sexless marriage?" Some are regretful: "I guess I just wanted to get married and have a baby, but I'm starting to look at everyone else and think they're happier than me." Some are wishful: "Anyone know if there's anything a woman can take to increase the sex drive. like herbal stuff or something?"

One night, a poster wrote in a panic that she'd found a receipt for lingerie that she'd never received. Over the next hour, dozens of anonymous posters walked her through the possibilities: Could it be a gift for her? Was he cheating? She posted again and again: She was a stay-at-home mom with small children; she usually contacted him by his cell phone. UrbanBaby posters argued tactics: Should she confront him when he got home? Coolly place the receipt on his plate? We eagerly waited for her to return and tell us what had happened, but she never came back, and no one could figure out whether that meant it was a fake or whether she had gotten bad news and never logged on again. Days after the incident, people continued posting to her, hoping she'd reappear, but she never did. And when someone would post claiming to be her, it was impossible to confirm if it was true.

Another night, on one of the numerous threads in which

people ask for people to post secrets, a woman revealed, "I may get married purposely to get divorced and get the alimony money." She was quickly called a sociopath. Posters were fascinated that she'd confess such a thing: How could she be so cold? And she did come across as cold but also intelligent and insightful, disassociated yet, like many of the posters on threads like this, startlingly self-reflective. She revealed that she prefers being a single mother. The guy is a nice-enough guy; he's just "there." And perhaps unsurprisingly, she was abandoned as a child and abused, although she told nobody about these experiences, or at least nobody except us.

Men on UrbanBaby are a very strange presence. They are necessary, and they are useless. They are critical, in both senses. Although women sometimes post loving accounts of their husbands' sexiness and smarts, more often men are less loving partners than objects for study: Did I know he was like this when I married him? Has he changed or have I? Search for *divorce* and you find women in all stages: about to leave, negotiating settlements, struggling with the mess of a custody dispute, and trying to keep things going.

12:13 a.m.
"we separated for a year—got back but it's still not great."
12:15 a.m.
"What's missing? What needs to be fixed?"
12:15 a.m.
"we have no sex life and we snap at each other."
12:16 a.m.
"maybe that is the reason."
12:16 a.m.
"yeah but we can't seem to snap out of it."

And then three simultaneous responses:

12:17 a.m.

"porn."

12:17 a.m.

"act."

12:17 a.m.

"pretend."

THE PROBLEM WITH NO NAME

In 1963, Betty Friedan called it "the Problem with No Name": the existential horror she'd discerned in the hearts of suburban housewives, the despair of the educated woman whose life had narrowed to the walls of her home. "As she made the beds, shopped for groceries, matched slipcover material, ate peanut butter sandwiches with her children, chauffeured Cub Scouts and Brownies, lay beside her husband at night—she was afraid to ask even of herself the silent question—'Is this all?'"

You'd think that that particular problem might have been solved by now. After all, these were emotions that were supposed to have bloomed in the dark, the product of female isolation: Every woman believed her sadness, her confusion, her inability to cope, was merely her own private neurosis. If she felt sleepy all the time or angry all the time; if she lay next to her husband feeling like an insane stranger to him— that was just because she was broken. Her natural femininity had soured like milk. And because no one could talk about it, it was hard to see the enormous hand of the culture that had created the isolated fingerprints of each mother's private breakdown.

But now the Problem with No Name has a million names (and almost as many anthologies) to describe it. In books like *Mommy Wars, Perfect Madness, The Mommy Myth,* and *The Bitch in the House,* authors have diagnosed the experience from every possible angle and offered as many contradictory solutions: Loosen up! Toughen up! Go to work! Stay home! Accept that this is just what men are like; refuse to accept that this is "just what men are like." Polemical performance artists like the writers Caitlin Flanagan and Ayelet Waldman (both UrbanBaby obsessions) use their own lives as fodder. Most recently, Linda Hirshman, author of *Get to Work: A Manifesto for Women of the World,* has joined the din, insisting women go back to the office at all costs, touting Friedan as her role model.

Hirshman has predictably caused her own ripples on UrbanBaby, especially among moms who resent her thesis that they are holding women back by staying home with their kids: "This Linda Hirshman chick is one nasty, bitter freak," reads one typical post. Yet her suggestion that we reread Friedan is a useful one, if only because going back to the Problem with No Name means experiencing waves of unsettling déjà vu. Because large swaths of Friedan's writing could have been written today, in response to Urban-Baby.

Here are some of the symptoms Friedan saw in the culture back in 1963: "In a New York hospital, a woman had a nervous breakdown when she found she could not breast-feed her baby." Women of all ages were desperate for marriage, defining themselves only by their association with men, "moving from one political club to another, taking evening courses in accounting or sailing, learning to play golf or ski, joining a number of churches in succession, going to bars alone, in their ceaseless search for a man." They were "taking tranquilizers like cough drops. You wake up in the morning, and you feel as if there's no point in going on another day like this. So you take a tranquilizer because it makes you not care so much that it's pointless."

All this free-floating anxiety was, Friedan noted with alarm, affecting the children. "Strange new problems are being reported in the growing generations of children whose mothers were always there, driving them around, helping them with their homework—an inability to endure pain or discipline or pursue any self-sustained goal of any sort, a devastating boredom with life. Educators are increasingly uneasy about the dependence, the lack of self-reliance, of the boys and girls who are entering college today." According to Friedan, a conference in the White House was even called on the subject, to discuss "the physical and muscular deterioration of American children: were they being over-nurtured? Sociologists noted the astounding organization of suburban children's lives: the lessons, parties, entertainments, play and study groups organized for them."

Maybe nothing has changed since 1963. Maybe everything has changed, and only the anxieties remain the same, with new labels pasted on top of the old pathologies—every distant "refrigerator mom" replaced by an overanxious "helicopter mom." Look closely and the new diagnoses begin to seem like the inverse of the old ones, just another way of pointing out that there is something wrong with mothers, at once neglectful and overprotective. As a culture, we seem perpetually afraid that there is something wrong with our children—that they are spoiled and weak and incapable of growing up.

But if you read UrbanBaby, it's hard not to be unsettled by the same conclusion that hit Friedan when she surveyed the mothers of America: that what seem like women's private struggles can be seen as an expression of their shared experience. Without a baby, it's easy to maintain the idea that you and your husband are atomic individuals, mavericks who shape your own fates; afterward, that notion slips away. In the 1970s, the solution for all this was supposed to be consciousness-raising groups. Women would gather together and spill their beans and then look for a pattern among the beans that had been spilled. These meetings weren't supposed to be merely bitch sessions or therapy. They were supposed to be an opportunity to look for the connections: to figure out how to restructure the world so that there was more room for everyone.

UrbanBaby and the Web in general can't offer that. They are too chaotic and too ephemeral; it's impossible to work for change when no one can agree on what needs to be changed, when even your closest ideological allies are nameless and disappear at 2 a.m.

But for all the bile on UrbanBaby, there's still something affecting and even powerful about seeing women's experiences spilled out so freely. In even the saddest exchanges, there's a feeling of community—the possibility for isolated women to hack their way through to a bigger picture. In these moments, women seem to be offering one another a way to reimagine the mother they've become.

10:56 p.m.
"Feeling sad about marriage. Have beautiful 4 mo son and a little bored with DH. We're not having sex that much, and I'm exhausted being home with ds. Money is tight and we aren't always kind when speaking to each other—rather sarcastic in fact."
10:58 p.m.
"it will get better. the first six months with a new baby can be very hard on a marriage."
10:58 p.m.
"you are me. i feel stuck."
11:02 p.m.
"4 mos we were still adjusting . . . and I felt like a single mom . . . and he didn't appreciate

how hard going back to work was (my choice, but
still hard), things started to get better soon
after"

11:04 p.m.

"It takes awhile to get used to a new person in
your home."

The Onion

iTunes to Sell You Your Home Videos for $1.99 Each

CUPERTINO, CALIFORNIA—Apple Computer, producer of the successful iPod MP3 player, is now offering consumers limited rights to buy their own home movies from the media store iTunes for $1.99 each.

"Ladies and gentlemen, the future of home-video viewing is now," Apple CEO Steve Jobs said at a media event Tuesday morning. "As soon as you record that precious footage of your daughter's first steps, you'll be able to buy it right back from iTunes and download it directly to your computer and video iPod."

Jobs emphasized that the videos will be presented unedited and in their original form, save for a small Apple logo in the lower right-hand corner of the image to protect the company's copyrighted materials from Internet piracy.

Added Jobs: "No more searching through your movies folder for that footage of your fiftieth wedding anniversary. Now all you need is a 768Kbps broadband connection and your credit card, and every timeless personal memory you've ever shot will be right at your fingertips."

"Apple has always been about access," said *MacAddict* editor Ian Smythe. "Thanks to this revolutionary new software, all your clips—from your son's bris to your father's dying message—are available to you, your loved ones, and the 20 million iTunes users, who will be able to view them on up to five different computers."

Apple currently owns an average of 20 gigabytes of digital footage per American family, and it has also acquired an enormous library of the tens of millions of analog-format home movies dating from the early decades of the twentieth century through 2001.

"No more disappointment for Cynthia Hamill of Hartford, Connecticut, when she realizes she can't find that tape of herself singing 'Sweet Caroline' in the bath as an eight-year-old," Jobs said. "For only a couple of bucks, that cherished moment can again be hers."

Early reaction to the home-video downloads has been positive. "$1.99 seems reasonable to be able to relive my high-school graduation anytime I want," said Patrick Boyd of Pensacola, Florida. "My parents don't understand the technology, but I can help them get it running whenever they want to watch it."

"It's just a matter of convenience," Mansfield, Ohio, resident Samantha Davidoff said. "Why should I sift through the dozens of unlabeled DV tapes in my closet to find that submission tape I made for *Extreme Makeover* when I can just do a search on iTunes? Repurchasing my own stuff has never been this intuitive."

However, some early users report running into technical glitches with the software.

"I was really looking forward to watching my son's Easter greeting from Iraq," Eugene, Oregon, resident Luka Bartoli said. "But the image froze and an alert came up saying it was temporarily unavailable due to low bandwidth. I miss my boy so much."

Some users say they have had trouble with the automated process by which previews are chosen for their new footage.

"We were all excited to watch [daughter] Tabitha's birth when we got home from the hospital, but we could only view a 30-second clip before we had to buy it," Harvey Gaddis of Tulsa, Oklahoma, said. "All we could see in the preview were some of the initial contractions."

Others say the pricing can be restrictive and is not always timely.

"I wanted to show my boyfriend a video I made for his birthday of me dancing in my underwear to our favorite song," Jessica Dupree of Manchester, New Hampshire, said. "But his credit card was declined. I guess he'll just have to get it from someone at work."

Despite these limitations, observers predict consumers will have little choice once they realize how vast and comprehensive the collection is. Many amateur filmmakers are already making a strong showing in the iTunes videos charts.

Eliza Quintana of Montclair, New Jersey, went online to purchase her daughter's fourth birthday party to find that it had reached No. 5 on the top video downloads.

Said Quintana: "I guess I'm not the only one who thinks she's the most adorable little girl in the world!"

Scan This Book!

What will happen to books? Reader, take heart!
Publisher, be very, very afraid. Internet search
engines will set them free. A manifesto.

In several dozen nondescript office buildings around the world, thousands of hourly workers bend over table-top scanners and haul dusty books into high-tech scanning booths. They are assembling the universal library page by page.

The dream is an old one: to have in one place all knowledge, past and present. All books, all documents, all conceptual works, in all languages. It is a familiar hope, in part because long ago we briefly built such a library. The great library at Alexandria, constructed around 300 B.C., was designed to hold all the scrolls circulating in the known world. At one time or another, the library held about half a million scrolls, estimated to have been between 30 and 70 percent of all books in existence then. But even before this great library was lost, the moment when all knowledge could be housed in a single building had passed. Since then, the constant expansion of information has overwhelmed our capacity to contain it. For 2,000 years, the universal library, together with other perennial longings like invisibility cloaks, antigravity shoes, and paperless offices, has been a mythical dream that kept receding further into the infinite future.

Until now. When Google announced in December 2004 that it would digitally scan the books of five major research libraries to make their contents searchable, the promise of a universal library was resurrected. Indeed, the explosive rise of the Web, going from nothing to everything in one decade, has encouraged us to believe in the impossible again. Might the long-heralded great library of all knowledge really be within our grasp?

Brewster Kahle, an archivist overseeing another scanning project, says that the universal library is now within reach. "This is our chance to one-up the Greeks!" he shouts. "It is really possible with the technology of today, not tomorrow. We can provide all the works of humankind to all the people of the world. It will be an achievement remembered for all time, like putting a man on the moon." And unlike the libraries of old, which were restricted to the elite, this library would be truly democratic, offering every book to every person.

But the technology that will bring us a planetary source of all written material will also, in the same gesture, transform the nature of what we now call the book and the libraries that hold them. The universal library and its "books" will be unlike any library or books we have known. Pushing us rapidly toward that Eden of everything, and away from the paradigm of the physical paper tome, is the hot technology of the search engine.

SCANNING THE LIBRARY OF LIBRARIES

Scanning technology has been around for decades, but digitized books didn't make much sense until recently, when search engines like Google, Yahoo, Ask, and MSN came along. When millions of books have been scanned and their texts are made available in a single database, search technology will enable us to grab and read any book ever written. Ideally, in such a complete library we should also be able to read any article ever written in any newspaper, magazine, or journal. And why stop there? The universal library should include a copy of every painting, photograph, film, and piece of music produced by all artists, present and past. Still more, it should include all radio and television broadcasts. Commercials too. And how can we forget the Web? The grand library naturally needs a copy of the billions of dead Web pages no longer online and the tens of millions of blog posts now gone—the ephemeral literature of our time. In short, the entire works of humankind, from the beginning of recorded history, in all languages, available to all people, all the time.

This is a very big library. But because of digital technology, you'll be able to reach inside it from almost any device that sports a screen. From the days of Sumerian clay tablets till now, humans have "published" at least 32 million books; 750 million articles and essays; 25 million songs; 500 million images; 500,000 movies; 3 million videos, TV shows, and short films; and 100 billion public Web pages. All this material is currently contained in all the libraries and archives of the world. When fully digitized, the whole lot could be compressed (at current technological rates) onto 50 petabyte hard disks. Today you need a building about the size of a small-town library to house 50 petabytes. With tomorrow's technology, it will all fit onto your iPod. When that happens, the library of all libraries will ride in your purse or wallet— if it doesn't plug directly into your brain with thin white cords. Some people alive today are surely hoping that they die before such things happen, and others, mostly the young, want to know what's taking so long. (Could we get it up and running by next week? They have a history project due.)

Technology accelerates the migration of all we know into the universal form of digital bits. Nikon will soon quit making film cameras for consumers, and Minolta already has: better think digital photos from now on. Nearly 100 percent of all contemporary recorded music has already been digitized, much of it by fans. About one-tenth of the 500,000 or so movies listed on the Internet Movie Database are now digitized on DVD. But because of copyright issues and the physical fact of the need to turn pages, the digitization of books has proceeded at a relative crawl. At most, one book in 20 has moved from analog to digital. So far, the universal library is a library without many books.

But that is changing very fast. Corporations and libraries around the world are now scanning about a million books per year. Amazon has digitized several hundred thousand contemporary books. In the heart of Silicon Valley, Stanford University (one of the five libraries collaborating with Google) is scanning its 8-million-book collection using a state-of-the art robot from the Swiss company 4DigitalBooks. This machine, the size of a small SUV, automatically turns the pages of each book as it scans it, at the rate of 1,000 pages per hour. A human operator places a book in a flat carriage, and then pneumatic robot fingers flip the pages—delicately enough to handle rare volumes—under the scanning eyes of digital cameras.

Like many other functions in our global economy, however, the real work has been happening far away, while we sleep. We are outsourcing the scanning of the universal library. Superstar, an entrepreneurial company based in Beijing, has scanned every book from 200 libraries in China. It has already digitized 1.3 million unique titles in Chinese, which it estimates is about half of all the books published in the Chinese language since 1949. It costs $30 to scan a book at Stanford but only $10 in China.

Raj Reddy, a professor at Carnegie Mellon University, decided to move a fair-size English-language library to where the cheap subsidized scanners were. In 2004, he borrowed 30,000 volumes from the storage rooms of the Carnegie Mellon library and the Carnegie Library and packed them off to China in a single shipping container to be scanned by an assembly line of workers paid by the Chinese. His project, which he calls the Million Book Project, is churning out 100,000 pages per day at 20 scanning stations in India and China. Reddy hopes to reach a million digitized books in two years.

The idea is to seed the bookless developing world with easily available texts. Superstar sells copies of books it scans back to the same university libraries it scans from. A university can expand a typical 60,000-volume library into a 1.3 million-volume one overnight. At about 50 cents per digital book acquired, it's a cheap way for a library to increase its collection. Bill McCoy, the general manager of Adobe's e-publishing business, says, "Some of us have thousands of books at home, can walk to wonderful big-box bookstores and well-stocked libraries, and can get Amazon.com to deliver next day. The most dramatic effect of digital libraries will be not on us, the well-booked, but on the billions of people worldwide who are underserved by ordinary paper books." It is these underbooked—students in Mali, scientists in Kazakhstan, elderly people in Peru—whose lives will be transformed when even the simplest unadorned version of the universal library is placed in their hands.

WHAT HAPPENS WHEN BOOKS CONNECT

The least important, but most discussed, aspects of digital reading have been these contentious questions: Will we give up the highly evolved technology of ink on paper and instead read on cumbersome machines? Or will we keep reading our paperbacks on the beach? For now, the answer is yes to both. Yes, publishers have lost millions of dollars on the long-prophesied e-book revolution that never occurred, while the number of physical books sold in the world each year continues to grow. At the same time, there are already more than half a billion PDF documents on the Web that people happily read on computers without printing them out, and still more people now spend hours watching movies on microscopic cell phone screens. The arsenal of our current display technology—from handheld gizmos to large flat screens—is already good enough to move books to their next stage of evolution: a full digital scan.

Yet the common vision of the library's future (even the e-book future) assumes that books will remain isolated items, independent from one another, just as they are on shelves in your public library. There, each book is pretty much unaware of the ones next to it. When an author completes a work, it is fixed and finished. Its only movement comes when a reader picks it up to animate it with his or her imagination. In this vision, the main advantage of the coming digital library is portability—the nifty translation of a book's full text into bits, which permits it to be read on a screen anywhere. But this vision misses the chief revolution birthed by scanning books: in the universal library, no book will be an island.

Turning inked letters into electronic dots that can be read on a screen is simply the first essential step in creating this new library. The real magic will come in the second act, as each word in each book is cross-linked, clustered, cited, extracted, indexed, analyzed, annotated, remixed, reassembled, and woven deeper into the culture than ever before. In the new world of books, every bit informs another; every page reads all the other pages.

In recent years, hundreds of thousands of enthusiastic amateurs have written and cross-referenced an entire online encyclopedia called Wikipedia. Buoyed by this success, many nerds believe that a billion readers can reliably weave together the pages of old books, one hyperlink at a time. Those with a passion for a special subject, obscure author, or favorite book will, over time, link up its important parts. Multiply that simple generous act by millions of readers, and the universal library can be integrated in full, by fans for fans.

In addition to a link, which explicitly connects one word or sentence or book to another, readers will also be able to add tags, a recent innovation on the Web but already a popular one. A tag is a public annotation, like a keyword or category name, that is hung on a file, page, picture, or song, enabling anyone to search for that file. For instance, on the photo-sharing site Flickr, hundreds of viewers will "tag" a photo submitted by another user with their own simple classifications of what they think the picture is about: "goat," "Paris," "goofy," "beach party." Because tags are user generated, when they move to the realm of books, they will be assigned faster, range wider, and serve better than out-of-date schemes like the Dewey decimal system, particularly in frontier or fringe areas like nanotechnology or body modification.

The link and the tag may be two of the most important inventions of the last 50 years. They get their initial wave of power when we first code them into bits of text, but their real transformative energies fire up as ordinary users click on them in the course of everyday Web surfing, unaware that each humdrum click "votes" on a link, elevating its rank of relevance. You may think you are just browsing, casually inspecting this paragraph or that page, but in fact you are anonymously marking up the Web with bread crumbs of attention. These bits of interest are gathered and analyzed by search engines in order to strengthen the relationship between the end points of every link and the connections suggested by each tag. This is a type of intelligence common on the Web but previously foreign to the world of books.

Once a book has been integrated into the new expanded library by means of this linking, its text will no longer be separate from the text in other books. For instance, today a serious nonfiction book will usually have a bibliography and some kind of footnotes. When books are deeply linked, you'll be able to click on the title in any bibliography or any footnote and find the actual book referred to in the footnote. The books referenced in that book's bibliography will themselves be available, and so you can hop through the library in the same way we hop through Web links, traveling from footnote to footnote to footnote until you reach the bottom of things.

Next come the words. Just as a Web article on, say, aquariums, can have some of its words linked to definitions of fish terms, any and all words in a digitized book can be hyperlinked to other parts of other books. Books, including fiction, will become a web of names and a community of ideas.

Search engines are transforming our culture because they harness the power of relationships, which is all links really are. There are about 100 billion Web pages, and each page holds, on average, 10 links. That's a trillion electrified connections coursing through the Web. This tangle of relationships is precisely what gives the Web its immense force. The static world of book knowledge is about to be transformed by the same elevation of relationships, as each page in a book discovers other pages and other books. Once text is digital, books seep out of their bindings and weave themselves together. The collective intelligence of a library allows us to see things we can't see in a single, isolated book.

When books are digitized, reading becomes a community activity. Bookmarks can be shared with fellow readers. Marginalia can be broadcast. Bibliographies swapped. You might get an alert that your friend Carl has annotated a favorite book of yours. A moment later, his links are yours. In a curious way, the universal library becomes one very, very, very large single text: the world's only book.

BOOKS: THE LIQUID VERSION

At the same time, once digitized, books can be unraveled into single pages or be reduced further, into snippets of a page. These snippets will be remixed into reordered books and virtual bookshelves. Just as the music audience now juggles and reorders songs into new albums (or "playlists," as they are called in iTunes), the universal library will encourage the creation of virtual "bookshelves"—a collection of texts, some as short as a paragraph, others as long as entire books, that form a library shelf's worth of specialized information. And as with music playlists, once created, these "bookshelves" will be published and swapped in the public commons. Indeed, some authors will begin to write books to be read as snippets or to be remixed as pages. The ability to purchase, read, and manipulate individual pages or sections is surely what will drive reference books (cookbooks, how-to manuals, travel guides) in the future. You might concoct your own "cookbook shelf" of Cajun recipes compiled from many different sources; it would include Web pages, magazine clippings, and entire Cajun cookbooks. Amazon currently offers you a chance to publish your own bookshelves (Amazon calls them "listmanias") as annotated lists of books you want to recommend on a particular esoteric subject.

And readers are already using Google Book Search to round up minilibraries on a certain topic—all books about Sweden, for instance, or books on clocks. Once snippets, articles, and pages of books become ubiquitous, shuffle-able, and transferable, users will earn prestige and perhaps income for curating an excellent collection.

Libraries (as well as many individuals) aren't eager to relinquish ink-on-paper editions, because the printed book is by far the most durable and reliable backup technology we have. Printed books require no mediating device to read and thus are immune to technological obsolescence. Paper is also extremely stable, compared with, say, hard drives or even CDs. In this way, the stability and fixity of a bound book are a blessing. It sits there unchanging, true to its original creation. But it sits alone.

So what happens when all the books in the world become a single liquid fabric of interconnected words and ideas? Four things: First, works on the margins of popularity will find a small audience larger than the near-zero audience they usually have now. Far out in the Long Tail of the distribution curve—that extended place of low-to-no sales where most of the books in the world live—digital interlinking will lift the readership of almost any title, no matter how esoteric. Second, the universal library will deepen our grasp of history, as every original document in the course of civilization is scanned and cross-linked. Third, the universal library of all books will cultivate a new sense of authority. If you can truly incorporate all texts—past and present, multilingual— on a particular subject, then you can have a clearer sense of what we as a civilization, a species, do know and don't know. The white spaces of our collective ignorance are highlighted, while the golden peaks of our knowledge are drawn with completeness. This degree of authority is only rarely achieved in scholarship today, but it will become routine.

Finally, the full, complete universal library of all works becomes more than just a better Ask Jeeves. Search on the Web becomes a new infrastructure for entirely new functions and services. Right now, if you mash up Google Maps and Monster.com, you get maps of where jobs are located by salary. In the same way, it is easy to see that in the great library, everything that has ever been written about, for example, Trafalgar Square in London could be present on that spot via a screen. In the same way, every object, event, or location on Earth would "know" everything that has ever been written about it in any book, in any language, at any time. From this deep structuring of knowledge comes a new culture of interaction and participation.

The main drawback of this vision is a big one. So far, the universal library lacks books. Despite the best efforts of bloggers and the creators of the Wikipedia, most of the world's expertise still resides in books. And a universal library without the contents of books is no universal library at all.

There are dozens of excellent reasons that books should quickly be made part of the emerging Web. But so far they have not been, at least not in great numbers. And there is only one reason: the hegemony of the copy.

THE TRIUMPH OF THE COPY

The desire of all creators is for their works to find their way into all minds. A text, a melody, a picture, or a story succeeds best if it is connected to as many ideas and other works as possible. Ideally, over time a work becomes so entangled in a culture that it appears to be inseparable from it, in the way that the Bible, Shakespeare's plays, "Cinderella," and the Mona Lisa are inseparable from ours. This tendency for creative ideas to infiltrate other works is great news for culture. In fact, this commingling of creations is culture.

In preindustrial times, exact copies of a work were rare for a simple reason: it was much easier to make your own version of a creation than to duplicate someone else's exactly. The amount of energy and attention needed to copy a scroll exactly, word for word, or to replicate a painting stroke by stroke exceeded the cost of paraphrasing it in your own style. So most works were altered, and often improved, by the borrower before they were passed on. Fairy tales evolved mythic depth as many different authors worked on them and as they migrated from spoken tales to other media (theater, music, painting). This system worked well for audiences and performers, but the only way for most creators to earn a living from their works was through the support of patrons.

That ancient economics of creation was overturned at the dawn of the industrial age by the technologies of mass production. Suddenly, the cost of duplication was lower than the cost of appropriation. With the advent of the printing press, it was now cheaper to print thousands of exact copies of a manuscript than to alter one by hand. Copy makers could profit more than creators. This imbalance led to the technology of copyright, which established a new order. Copyright bestowed upon the creator of a work a temporary monopoly—for 14 years, in the United States—over any copies of the work. The idea was to encourage authors and artists to create yet more works that could be cheaply copied and thus fill the culture with public works.

Not coincidentally, public libraries first began to flourish with the advent of cheap copies. Before the industrial age, libraries were primarily the property of the wealthy elite. With mass production, every small town could afford to put duplicates of the greatest works of humanity on wooden shelves in the village square. Mass access to public-library books inspired scholarship, reviewing, and education, activities exempted in part from the monopoly of copyright in the United States because they moved creative works toward the public commons sooner, weaving them into the fabric of common culture while still remaining under the author's copyright. These are now known as "fair uses."

This wonderful balance was undone by good intentions. The first was a new copyright law passed by Congress in 1976. According to the new law, creators no longer had to register or renew copyright; the simple act of creating something bestowed it with instant and automatic rights. By default, each new work was born under private ownership rather than in the public commons. At first, this reversal seemed to serve the culture of creation well. All works that could be copied gained instant and deep ownership, and artists and authors were happy. But the 1976 law, and various revisions and extensions that followed it, made it extremely difficult to move a work into the public commons, where human creations naturally belong and were originally intended to reside. As more intellectual property became owned by corporations rather than by individuals, those corporations successfully lobbied Congress to keep extending the once-brief protection enabled by copyright in order to prevent works from returning to the public domain. With constant nudging, Congress moved the expiration date from 14 years to 28 to 42 and then to 56.

While corporations and legislators were moving the goalposts back, technology was accelerating forward. In Internet time, even 14 years is a long time for a monopoly; a monopoly that lasts a human lifetime is essentially an eternity. So when Congress voted in 1998 to extend copyright an additional 70 years beyond the life span of a creator—to a point where it could not possibly serve its original purpose as an incentive to keep that creator working—it was obvious to all that copyright now existed primarily to protect a threatened business model. And because Congress at the same time tacked a 20-year extension onto all existing copyrights, nothing—no published creative works of any type—will fall out of protection and return to the public domain until 2019. Almost everything created today will not return to the commons until the next century. Thus the stream of shared material that anyone can improve (think "A Thousand and One Nights" or "Amazing Grace" or "Beauty and the Beast") will largely dry up.

In the world of books, the indefinite extension of copyright has had a perverse effect. It has created a vast collection of works that have been abandoned by publishers, a continent of books left permanently in the dark. In most cases, the original publisher simply doesn't find it profitable to keep these books in print. In other cases, the publishing company doesn't know whether it even owns the work, since author contracts in the past were not as explicit as they are now. The size of this abandoned library is shocking: about 75 percent of all books in the world's libraries are orphaned. Only about 15 percent of all books are in the public domain. A luckier 10 percent are still in print. The rest, the bulk of our universal library, is dark.

THE MORAL IMPERATIVE TO SCAN

The 15 percent of the world's 32 million cataloged books that are in the public domain are freely available for anyone to borrow, imitate, publish, or copy wholesale. Almost the entire current scanning effort by American libraries is aimed at this 15 percent. The Million Book Project mines this small sliver of the pie, as does Google. Because they are in the commons, no law hinders this 15 percent from being scanned and added to the universal library.

The approximately 10 percent of all books actively in

print will also be scanned before long. Amazon carries at least 4 million books, which includes multiple editions of the same title. Amazon is slowly scanning all of them. Recently, several big American publishers have declared themselves eager to move their entire backlist of books into the digital sphere. Many of them are working with Google in a partnership program in which Google scans their books, offers sample pages (controlled by the publisher) to readers, and points readers to where they can buy the actual book. No one doubts electronic books will make money eventually. Simple commercial incentives guarantee that all in-print and backlisted books will before long be scanned into the great library. That's not the problem.

The major problem for large publishers is that they are not certain what they actually own. If you would like to amuse yourself, pick an out-of-print book from the library and try to determine who owns its copyright. It's not easy. There is no list of copyrighted works. The Library of Congress does not have a catalog. The publishers don't have an exhaustive list, not even of their own imprints (though they say they are working on it). The older, the more obscure the work, the less likely a publisher will be able to tell you (that is, if the publisher still exists) whether the copyright has reverted to the author, whether the author is alive or dead, whether the copyright has been sold to another company, whether the publisher still owns the copyright, or whether it plans to resurrect or scan it. Plan on having a lot of spare time and patience if you inquire. I recently spent two years trying to track down the copyright to a book that led me to Random House. Does the company own it? Can I reproduce it? Three years later, the company is still working on its answer. The prospect of tracking down the copyright— with any certainty—of the roughly 25 million orphaned books is simply ludicrous.

Which leaves 75 percent of the known texts of humans in the dark. The legal limbo surrounding their status as copies prevents them from being digitized. No one argues that these are all masterpieces, but there is history and context enough in their pages to not let them disappear. And if they are not scanned, they in effect will disappear. But with copyright hyperextended beyond reason (the Supreme Court in 2003 declared the law dumb but not unconstitutional), none of this dark library will return to the public domain (and be cleared for scanning) until at least 2019. With no commercial incentive to entice uncertain publishers to pay for scanning these orphan works, they will vanish from view. According to Peter Brantley, director of technology for the California Digital Library, "We have a moral imperative to reach out to our library shelves, grab the material that is orphaned, and set it on top of scanners."

No one was able to unravel the Gordian knot of copydom until 2004, when Google came up with a clever solution. In addition to scanning the 15 percent out-of-copyright public-domain books with their library partners and the 10 percent in-print books with their publishing partners, Google executives declared that they would also scan the 75 percent out-of-print books that no one else would touch. They would scan the entire book, without resolving its legal status, which would allow the full text to be indexed on Google's internal computers and searched by anyone. But the company would show to readers only a few selected sentence-long snippets from the book at a time. Google's lawyers argued that the snippets the company was proposing were something like a quote or an excerpt in a review and thus should qualify as a "fair use."

Google's plan was to scan the full text of every book in five major libraries: the more than 10 million titles held by Stanford, Harvard, Oxford, the University of Michigan, and the New York Public Library. Every book would be indexed, but each would show up in search results in different ways. For out-of-copyright books, Google would show the whole book, page by page. For the in-print books, Google would work with publishers and let them decide what parts of their books would be shown and under what conditions. For the dark orphans, Google would show only limited snippets. And any copyright holder (author or corporation) who could establish ownership of a supposed orphan could ask Google to remove the snippets for any reason.

At first glance, it seemed genius. By scanning all books (something only Google had the cash to do), the company would advance its mission to organize all knowledge. It would let books be searchable, and it could potentially sell ads on those searches, although it does not do that currently. In the same stroke, Google would rescue the lost and forgotten 75 percent of the library. For many authors, this all-out campaign was a salvation. Google became a discovery tool, if not a marketing program. While a few best-selling authors fear piracy, every author fears obscurity. Enabling their works to be found in the same universal search box as everything else in the world was good news for authors and good news for an industry that needed some. For authors with books in the publisher program and for authors of books abandoned by a publisher, Google unleashed a chance that more people would at least read, and perhaps buy, the creation they had sweated for years to complete.

THE CASE AGAINST GOOGLE

Some authors and many publishers found more evil than genius in Google's plan. Two points outraged them: the virtual copy of the book that sat on Google's indexing server and Google's assumption that it could scan first and ask questions later. On both counts the authors and publishers accused Google of blatant copyright infringement. When negotiations failed last fall, the Authors Guild and five big publishing companies sued Google. Their argument was simple: Why shouldn't Google share its ad revenue (if any) with the copyright owners? And why shouldn't Google have to ask permission from the legal copyright holder before scanning the work in any case? (I have divided loyalties in the case. The current publisher of my books is suing Google to protect my earnings as an author. At the same time, I earn income from Google Adsense ads placed on my blog.)

One mark of the complexity of this issue is that the publishers suing were, and still are, committed partners in the Google Book Search Partner Program. They still want Google to index and search their in-print books, even when they are scanning the books themselves, because, they say, search is a discovery tool for readers. The ability to search the scans of all books is good for profits.

The argument about sharing revenue is not about the 3 or 4 million books that publishers care about and keep in print, because Google is sharing revenues for those books with publishers. (Google says publishers receive the "majority share" of the income from the small ads placed on partner-program pages.) The argument is about the 75 percent of books that have been abandoned by publishers as uneconomical. One curious fact, of course, is that publishers only care about these orphans now because Google has shifted the economic equation; because of Book Search, these dark books may now have some sparks in them, and the publishers don't want this potential revenue stream to slip away from them. They are now busy digging deep into their records to see what part of the darkness they can declare as their own.

The second complaint against Google is more complex. Google argues that it is nearly impossible to track down copyright holders of orphan works, and so, it says, it must scan those books first and only afterward honor any legitimate requests to remove the scan. In this way, Google follows the protocol of the Internet. Google scans all Web pages; if it's on the Web, it's scanned. Web pages, by default, are born copyrighted. Google, therefore, regularly copies billions of copyrighted pages into its index for the public to search. But if you don't want Google to search your Web site, you can stick some code on your home page with a no-searching sign, and Google and every other search engine will stay out. A Webmaster thus can opt out of search. (Few do.) Google applies the same principle of opting-out to Book Search. It is up to you as an author to notify Google if you don't want the company to scan or search your copyrighted material. This might be a reasonable approach for Google to demand from an author or publisher if Google were the only search company around. But search technology is becoming a commodity, and if it turns out there is any money in it, it is not impossible to imagine 100 mavericks scanning out-of-print books. Should you as a creator be obliged to find and notify each and every geek who scanned your work, if for some reason you did not want it indexed? What if you miss one?

There is a technical solution to this problem: for the search companies to compile and maintain a common list of no-scan copyright holders. A publisher or author who doesn't want a work scanned notifies the keepers of the common list once, and anyone conducting scanning would have to remove material that was listed. Since Google, like all the other big search companies—Microsoft, Amazon, and Yahoo—is foremost a technical-solution company, it favors this approach. But the battle never got that far.

WHEN BUSINESS MODELS COLLIDE

In thinking about the arguments around search, I realized that there are many ways to conceive of this conflict. At first, I thought that this was a misunderstanding between people of the book, who favor solutions by laws, and people of the screen, who favor technology as a solution to all problems. Last November, the New York Public Library (one of the "Google Five") sponsored a debate between representatives of authors and publishers and supporters of Google. I was tickled to see that up on the stage the defenders of the book were from the East Coast and the defenders of the screen were from the West Coast. But while it's true that there's a strand of cultural conflict here, I eventually settled on a different framework, one that I found more useful. This is a clash of business models.

Authors and publishers (including publishers of music and film) have relied for years on cheap mass-produced copies protected from counterfeits and pirates by a strong law based on the dominance of copies and on a public educated to respect the sanctity of a copy. This model has, in the last century or so, produced the greatest flowering of human achievement the world has ever seen, a magnificent golden age of creative works. Protected physical copies have enabled millions of people to earn a living directly from the sale of their art to the audience, without the weird dynamics of patronage. Not only did authors and artists benefit from this model, but the audience did, too. For the first time, billions of ordinary people were able to come in regular contact with a great work. In Mozart's day, few people ever heard one of his symphonies more than once. With the advent of cheap audio recordings, a barber in Java could listen to them all day long.

But a new regime of digital technology has now dis-

88 *Kevin Kelly*

rupted all business models based on mass-produced copies, including individual livelihoods of artists. The contours of the electronic economy are still emerging, but while they do, the wealth derived from the old business model is being spent to try to protect that old model, through legislation and enforcement. Laws based on the mass-produced copy artifact are being taken to the extreme, while desperate measures to outlaw new technologies in the marketplace "for our protection" are introduced in misguided righteousness. (This is to be expected. The fact is, entire industries and the fortunes of those working in them are threatened with demise. Newspapers and magazines, Hollywood, record labels, broadcasters, and many hard-working and wonderful creative people in those fields have to change the model of how they earn money. Not all will make it.)

The new model, of course, is based on the intangible assets of digital bits, where copies are no longer cheap but free. They freely flow everywhere. As computers retrieve images from the Web or display texts from a server, they make temporary internal copies of those works. In fact, every action you take on the Net or invoke on your computer requires a copy of something to be made. This peculiar superconductivity of copies spills out of the guts of computers into the culture of computers. Many methods have been employed to try to stop the indiscriminate spread of copies, including copy-protection schemes, hardware-crippling devices, education programs, even legislation, but all have proved ineffectual. The remedies are rejected by consumers and ignored by pirates.

As copies have been dethroned, the economic model built on them is collapsing. In a regime of superabundant free copies, copies lose value. They are no longer the basis of wealth. Now relationships, links, connection, and sharing are. Value has shifted away from a copy toward the many ways to recall, annotate, personalize, edit, authenticate, display, mark, transfer, and engage a work. Authors and artists can make (and have made) their livings selling aspects of their works other than inexpensive copies of them. They can sell performances, access to the creator, personalization, add-on information, the scarcity of attention (via ads), sponsorship, periodic subscriptions—in short, all the many values that cannot be copied. The cheap copy becomes the "discovery tool" that markets these other intangible valuables. But selling things-that-cannot-be-copied is far from ideal for many creative people. The new model is rife with problems (or opportunities). For one thing, the laws governing creating and rewarding creators still revolve around the now-fragile model of valuable copies.

SEARCH CHANGES EVERYTHING

The search-engine companies, including Google, operate in the new regime. Search is a wholly new concept, not foreseen in version 1.0 of our intellectual-property law. In the words of a recent ruling by the United States District Court for Nevada, search has a "transformative purpose," adding new social value to what it searches. What search uncovers is not just keywords but also the inherent value of connection. While almost every artist recognizes that the value of a creation ultimately rests in the value he or she personally gets from creating it (and for a few artists that value is sufficient), it is also true that the value of any work is increased the more it is shared. The technology of search maximizes the value of a creative work by allowing a billion new connections into it, often a billion new connections that were previously inconceivable. Things can be found by search only if they radiate potential connections. These potential relationships can be as simple as a title or as deep as hyperlinked footnotes that lead to active pages, which are also footnoted. It may be as straightforward as a song published intact or as complex as access to the individual instrument tracks—or even individual notes.

Search opens up creations. It promotes the civic nature of publishing. Having searchable works is good for culture. It is so good, in fact, that we can now state a new covenant: Copyrights must be counterbalanced by copyduties. In exchange for public protection of a work's copies (what we call copyright), a creator has an obligation to allow that work to be searched. No search, no copyright. As a song, movie, novel, or poem is searched, the potential connections it radiates seep into society in a much deeper way than the simple publication of a duplicated copy ever could.

We see this effect most clearly in science. Science is on a long-term campaign to bring all knowledge in the world into one vast, interconnected, footnoted, peer-reviewed web of facts. Independent facts, even those that make sense in their own world, are of little value to science. (The pseudo- and parasciences are nothing less, in fact, than small pools of knowledge that are not connected to the large network of science.) In this way, every new observation or bit of data brought into the web of science enhances the value of all other data points. In science, there is a natural duty to make what is known searchable. No one argues that scientists should be paid when someone finds or duplicates their results. Instead, we have devised other ways to compensate them for their vital work. They are rewarded for the degree that their work is cited, shared, linked, and connected in their publications, which they do not own. They are financed with extremely short-term (20-year) patent monopolies for their ideas, short enough to truly inspire them to invent more, sooner. To a large degree, they make their living by giving away copies of their intellectual property in one fashion or another.

The legal clash between the book copy and the searchable Web promises to be a long one. Jane Friedman, the CEO of HarperCollins, which is supporting the suit against Google (while remaining a publishing partner), declared, "I don't expect this suit to be resolved in my lifetime." She's right. The courts may haggle forever as this complex issue works its way to the top. In the end, it won't matter; technology will resolve this discontinuity first. The Chinese scanning factories, which operate under their own, looser intellectual-property assumptions, will keep churning out digital books. And as scanning technology becomes faster, better, and cheaper, fans may do what they did to music and simply digitize their own libraries.

What is the technology telling us? That copies don't count any more. Copies of isolated books, bound between inert covers, soon won't mean much. Copies of their texts, however, will gain in meaning as they multiply by the millions and are flung around the world, indexed and copied again. What counts are the ways in which these common copies of a creative work can be linked, manipulated, annotated, tagged, highlighted, bookmarked, translated, enlivened by other media, and sewn together into the universal library. Soon a book outside the library will be like a Web page outside the Web, gasping for air. Indeed, the only way for books to retain their waning authority in our culture is to wire their texts into the universal library.

But the reign of livelihoods based on the copy is not over. In the next few years, lobbyists for book publishers, movie studios, and record companies will exert every effort to mandate the extinction of the "indiscriminate flow of copies," even if it means outlawing better hardware. Too many creative people depend on the business model revolving around copies for it to pass quietly. For their benefit, copyright law will not change suddenly.

But it will adapt eventually. The reign of the copy is no match for the bias of technology. All new works will be born digital, and they will flow into the universal library as you might add more words to a long story. The great continent of orphan works, the 25 million older books born analog and caught between the law and users, will be scanned. Whether this vast mountain of dark books is scanned by Google, the Library of Congress, the Chinese, or readers themselves, it will be scanned well before its legal status is resolved simply because technology makes it so easy to do and so valuable when done. In the clash between the conventions of the book and the protocols of the screen, the screen will prevail. On this screen, now visible to 1 billion people on Earth, the technology of search will transform isolated books into the universal library of all human knowledge.

Clive Thompson

A Head for Detail

*Gordon Bell feeds every piece of his life into a sur-
rogate brain, and soon the rest of us will be able to
do the same. But does perfect memory make you
smarter, or just drive you nuts?*

Gordon Bell will never forget what I look like. He'll never
forget what I sound like, either. Actually, he'll never forget
a single detail about me.

That's because when I first met the affable 72-year-old
computer scientist at the offices of Microsoft Research Labs,
in Redmond, Washington, he was carefully recording my
every move. He had a tiny bug-eyed camera around his neck
and a small audio recorder at his elbow. As we chatted about
various topics—Australian jazz musicians, his futuristic cell
phone, the Seattle area's gorgeous weather—Bell's gear quietly logged my every gesture and all my blathering small
talk, snapping a picture every 60 seconds. Back at his office,
his computer had carefully archived every document related
to me: all the e-mail I'd sent him, copies of my articles he'd
read, pages he'd surfed on my blog.

"Oh, I've got everything," Bell said cheerily. And when
I saw him the next day, down in his cramped personal office
in San Francisco, he offered to give me a glimpse of the

memories he'd collected. He plunked down in front of his computer, pulled up a browser, typed in "Clive Fast Company," and there they were: Hundreds of pictures of the meeting scrolled by on his screen, and the sound of our day-old conversation filled the room. It was a deeply strange feeling. My random chitchat is being preserved? For all eternity? He nodded, pointing to a mundane Dell computer parked beneath his desk. His massive store of data. His "surrogate brain."

Because I'm not the only thing Gordon Bell will never forget. His goal is never to forget anything.

For the past seven years, Bell has been conducting an audacious experiment in "lifelogging"—creating a near-total digital record of his experience. His custom-designed software, "MyLifeBits," saves everything it can get its hands on. For every piece of e-mail he sends and receives, every document he types, every chat session he engages in, every Web page he surfs, a copy is scooped up and stashed away. MyLifeBits records his telephone calls and archives every picture—up to 1,000 a day—snapped by his automatic "SenseCam," that device slung around his neck. He has even stowed his entire past: The massive stacks of documents from his 47-year computer career, first as a millionaire executive, then as a government Internet bureaucrat, have been hoovered up and scanned in. The last time he counted, MyLifeBits had more than 101,000 e-mails, almost 15,000 Word and PDF documents, 99,000 Web pages, and 44,000 pictures.

"And that," he cackles, "is a s—tload of stuff."

That load has endowed Bell with the ability to perform supernatural feats of memory. He can dredge up the precise contents of an inspirational note above his desk 30 years ago (a set of aphorisms, including "Start many fires"). He knows who passed him on the street on the way to work four weeks ago. And when someone disputes his recollection of a conference call the previous day, he can end the argument by pulling up the audio stream and listening to it again. Instantly.

"It gives you kind of a feeling of cleanliness," Bell tells me. "I can offload my memory. I feel much freer about remembering something now. I've got this machine, this slave, that does it."

It gives his mind the chance, he says, to be more playful, to have more energy for creative thinking. But it is also a double-edged sword. Bell suspects MyLifeBits might be slowly degrading his real, carbon-based brain's ability to remember clearly. When you have an outboard mind doing the scut work, you tend to get out of practice. "It's like doing arithmetic," he says. "Who does it anymore? You've got pocket calculators for that. I know I can do long division. But I haven't done it for a long time."

It's a crazy experiment. But perhaps its craziest aspect is that soon you'll be part of it too—whether you want to be or not. The way Bell sees it, computers and the Internet are now rapidly becoming capable of storing everything you do and see. Hard-drive space has exploded in size, and every day people are recording more and more of their lives: We blog about our thoughts, upload personal pictures to Flickr, save every e-mail on our infinitely expanding Gmail accounts, shoot video on our cell phones, record phone calls straight to our hard drives when we use Skype.

"People say, 'Oh, what you're doing is revolutionary!'" Bell says. "I say, 'No, no, it's *evolutionary*. Because it's happening to you. It's happening as you speak.'"

So what will life be like when nothing is forgotten? Provocative as that question may be, it's hardly theoretical. The thinking behind MyLifeBits and other lifelogging research is already seeping into our lives. It's changing the way our search engines work. It's affecting corporate strategy. And the power of machines to create boundless memory—and to augment and even transform human thinking—is only going to become more pronounced. We've arrived at a time when the memory of machines creates ideas we've never considered.

PAPERLESS TRAIL

You could trace the notion of perfect recall back to 1945, when presidential science adviser Vannevar Bush published a provocative essay in the *Atlantic Monthly* entitled "As We May Think." Bush argued that man's mind could be perfected by technology. He envisioned a device called a Memex, "in which an individual stores all his books, records, and communications, and which is mechanized so that it may be consulted with exceeding speed and flexibility." A user would wear a "walnut-sized" camera on his forehead, capturing everything he saw, and then sit down at his Memex to browse thousands of personal letters, newspapers, and encyclopedias instantly. It would be, Bush argued, "an enlarged intimate supplement to his memory."

@ls:MyLifeBits was born from a much humbler idea: Bell was sick of carting around stacks of paper. He was a veteran of the computer revolution—indeed, he helped kick-start it in the 1960s and 1970s by building the first refrigerator-sized "minicomputers" for DEC, a pioneering computer firm. In the 1980s, he helped the government bootstrap the Internet into existence and then worked as an angel in Silicon Valley, growing wealthier and wealthier as his investments took flight. Hired in 1995 by Microsoft Research Labs, a wing of the company devoted to designing the future of computers, Bell was given carte blanche. He decided to become the first person in history "to truly go paperless."

So he bought a scanner, and his poor assistant, Vicki, a witty, motherly 56-year-old, began the arduous slog of making PDFs of four enormous filing cabinets' worth of stuff. The archive begins with photos of Bell's mother's birth in 1900 and basically never stops, sucking in everything from the sublime to the ridiculous: Bell's medical records, his Japanese-made notebooks filled with his elegant sketches of computer circuitry, phone bills, stickie notes, a copy of a "robot driver's license" he got a couple of years ago.

His appetite whetted, Bell decided to store even more data. So he turned to two Microsoft researchers, Jim Gemmell and Roger Lueder, who built software to automatically save digital copies of everything Bell generated: chat transcripts, every Web page he looked at, even records of his keystrokes. Then Lyndsay Williams, a Microsoft inventor in Cambridge, England, came up with an even more radical idea: the SenseCam, which creates a visual record of his life, like a personal security camera. It would snap pictures either at regular intervals or when triggered by a meaningful event, such as when its infrared detectors sensed someone standing in front of Bell or when its light sensors saw that he'd entered a new room. A tagalong GPS device stamps each picture with its geographic location.

At first, Bell was worried about filling up his hard-drive space too quickly. He accumulates one gig of information a month, and at that clip, the average MyLifeBits for a 72-year-old person would require one to three terabytes—a hefty amount of storage. But by 2000, driven by teenagers' insatiable desire to store MP3s and video clips, hard drives had dropped radically in price and grown enormously in capacity. Bell figures that in a few years, even a cheap cell phone will have enough space to store your entire existence. "We've gone from this period of scarcity, when you had to always go, 'Jeez, I can't keep this video file because my hard drive is full,' to the opposite," Bell says. "I tell people, 'Never throw anything out. You'll never have to worry about space for the rest of your life.'"

Slowly, in often subtle ways, MyLifeBits began to affect Bell's life. During a phone call to discuss a heart problem last year, Bell couldn't follow his doctor's flood of jargon— but he could listen to the call again and decode it at his leisure. A friend passed away; Bell was able to pluck a piece of 20-year-old correspondence from the mists for his eulogy. Meanwhile, the presence of the SenseCam and audio recorder began creeping out his "significant other," who wasn't sure she liked having everything set in stone. "We'd be talking, and she'd suddenly go, 'You didn't record that, did you?'" Bell chuckles. "And I'd admit, Yeah, I did. 'Delete it! Delete it!'"

Bell also discovered he was getting annoyed by experiences that couldn't be stuffed into a hard drive. During a ride in a cab in Australia, a tiny security cam surveyed him, and he wondered why he couldn't automatically get a copy of the feed. And books, in particular, drive him crazy. "I virtually refuse to own any books at this point," he complained at one point. "I mean, I get them, I look at them, I occasionally read them. But then I give them away, because they're not in my memory. To me they're almost gone."

TOTAL RECALL

If it's not in your database, it doesn't exist. That's the sort of eerie philosophical proposition Bell's project raises. He has a superhuman brain: Does that change the nature of being human?

For Bell, MyLifeBits has reduced the nagging anxiety we face every day at work. We meet an important stranger and panic about whether we'll remember her name and position. We browse through Web pages, wondering absently if we should take the extra few seconds to bookmark something for future reference. These tiny but draining bits of mental toil have fallen from Bell's cognitive load—a luxury even for me as I reported this piece: Copies of his old memos from DEC? The list of people he considered hiring a few years ago to head up Microsoft Research, with handwritten notes on their strengths and weaknesses? He had pristine copies at hand.

Martin Conway, a psychologist and memory expert at the University of Leeds, argues that projects like MyLifeBits can actually improve mental health by freeing our brains to be more productive and more creative. "We're moving into an age when technology is going to massively enhance our cognitive abilities, our problem-solving abilities," he says. It's rather like the way Google has already become an indispensable part of how people think about things—sitting at their desks, constantly tapping into the world's massive trove of information. "Your real memory becomes a sort of executive manager for all these other technological abilities."

"Forgetting is how we make sense of life," says one skeptic. "We need to forget."

Personal-productivity guru David Allen also has long argued that the frailty of everyday memory is the primary source of stress for overburdened corporate types. We sit around anxious about our to-do lists because we can never entirely remember them (while we're at work) or entirely forget them (when we're not).

Yet Bell's project has also made some observers nervous. It may not be a good idea, they argue, to tamper with human memory—because it's such a powerful part of what makes us who we are.

"I'm a big fan of forgetting," says Frank Nack, a German computer scientist who published a critique of lifelogging experiments last winter. "It's how we make sense of life, how we interpret things. Everybody is building a life story; we all need to forget certain stages. I don't want to be reminded of everything I said." Forgetting, he points out, is key to cultural concepts like forgiveness and nostalgia. Sure, we lose track of most of what happens to us—but that natural filtering process results in what we call knowledge and wisdom. When memories are only a click away, Nack says, they're cheapened. Without the difficult act of pulling something from the crannies of the mind, we become like the hapless high-school student who gets 2 million hits for a search on "World War II" and has no way of prioritizing them.

James L. McGaugh, a memory expert at the University of California, Irvine, points to the sad spectacle of Funes, a character in a Jorge Luis Borges story who suffers a head injury that renders him incapable of forgetting. "He says, 'My mind is like a garbage heap.' That's what it'd be like," McGaugh adds. "You have to watch what you wish for with memory."

As Bell's significant other realized, if everyone had a record of every conversation, it could turn everyday life and work into a maddening series of *gotchas*. Imagine that prig in your weekly meeting confronting you with an ill-advised comment you made three months ago. (On the other hand, imagine having a handy record of your boss's promises about when, precisely, he'd get that report back to you.) College graduates are already getting a taste of life in a world of persistent memory. Last spring, many found themselves getting turned down for entry-level jobs after prospective employers Googled them and unearthed tales of debauchery—with photos!—on their MySpace pages. Some corporations already erase all e-mail older than a few months for fear of suffering the fate of Enron or Microsoft, companies that had humiliating years-old correspondence subpoenaed.

What's more, knowing that everything is being logged might actually turn us into different people. We might be less flamboyant, less funny, less willing to say risky but potentially useful things, much as politicians on-camera tamp down their public statements into stifled happytalk. "There'd be a chilling effect," particularly early on, says Mark Federman, former strategist for the McLuhan Program in Culture and Technology, a high-tech think tank. "We'd all be on our best behavior. Reality would become reality TV."

As for Bell, he acknowledges that all this remembering could have a downside. "Fifty years from now, do you want to know that, gee, I visited a porn site today?" he asks with a smile. And when it comes to corporate information, he admits that "ownership, deniability, privacy, expungability—they're important." Microsoft hasn't yet objected to all of its sensitive corporate memos going into Bell's off-site brain, but he suspects that day will come. When he eventually retires, he says, he'll be in the weird position of having to shave those memories off and give them back. "I'll need a lobotomy," he says, only half joking.

Still, Bell insists the trend toward total memory isn't going away. More and more, it is happening automatically. Those tens of millions of bloggers and Flickr users—all out there recording their thoughts and pictures—have clearly decided that there's enormous value not just in capturing those thoughts but in sharing them with the public. The choice isn't whether you'll join the revolution but whether you'll embrace it.

SHAPING CHAOS

For all of its machine muscle, Bell's virtual memory wasn't quite what I imagined. When I first heard about his work, I expected someone who would dazzle me mentally, pulling off feats of recollection like some cyborgian savant: *Quick— give me the name of a random seatmate on a flight last July! List all the ingredients from that stroganoff you made back in college!* And indeed, hanging out with a mnemonist could be quite unsettling. One day, Bell was trying to describe to me a superb jazz-trumpet performance he'd attended in Australia the week before and then realized he didn't have to— he just found the audio file of the event and played it, the sinuous solo blasting out his computer speakers. (He concedes, sheepishly, that it probably wasn't quite legal to record the event.)

MyLifeBits is now so big that it faces a classic problem of information management: It's hellishly difficult to search, and Bell often finds himself lost in the forest. He hunts for an e-mail but can't lay his hands on it. He gropes for a document, but it eludes him. While eating lunch in San Francisco, he tells me about a Paul Krugman column he liked, so I ask him to show it to me. But it's like pulling teeth: A MyLifeBits search for "Paul Krugman" produces scores of columns, and Bell can't quite filter out the right one. When I ask him to locate a phone call from one of his colleagues, he hits a bug: He can locate the name of the file, but when he clicks on it the data are AWOL. "Where the hell is this friggin' phone call?" he mutters to himself, pecking at the keyboard. "I either get nothing, or I get too much!"

Granted, MyLifeBits is an experimental demo and thus naturally unstable. And even when his system is failing, Bell remains pretty bemused about everything, displaying the perpetual geniality of all brilliant, accomplished, wealthy older men who've long ceased to care what anyone thinks of them. Still, as I watch the hunt for the missing call, it feels like some creepy sci-fi version of Alzheimer's or a scene plucked out of a bleak Philip K. Dick novel: Our antihero has an external brain with perfect recall, but it's locked up tight and he can't get in—a cyborg estranged from his own limbs.

This turns out to be the central question behind MyLifeBits: Yes, it's possible to store a lifetime of memories, but what do you do with them?

To figure that out, I made a visit to Mary Czerwinski, a principal research scientist at Microsoft Research Labs whose team has developed "Facetmap," an audacious piece of software designed to visualize the contents of Bell's cyber-memory.

When I meet the energetic, hyperverbal Czerwinski, she pulls me over to a massive three-foot-by-three-foot LCD monitor on her office wall. On-screen there's a collection of colorful blobs representing different parts of Bell's life. There's a blob for people, another for calendar dates, and a bunch for different types of documents like e-mail or Word files. She shows me how it works: If you click on any blob, it instantly expands to show you everything it contains. Click on the blob for "Jim Gemmell," Bell's main collaborator, and you'll see a blob containing all their e-mail traffic, another with documents that mention Gemmell's name, and a third with events where he appears. The more data in each category, the bigger the blob, "so you can quickly see which area has had the most action," she notes.

But the truly intriguing part about Facetmap is that it shows how Bell's information is connected. I start poking around, clicking on Gemmell's blob and then drilling down to a particular e-mail Bell sent him on February 25, 2005. As I zoom in, the software automatically creates new blobs showing everything else Bell did on that day—e-mails to other colleagues, photos Bell took, Web sites he viewed. It feels like flying freely through Bell's memories, flitting anywhere I want. And it re-creates those same loosey-goosey linkages that tie memories together in our real-life minds. Were this my own computer, I could zip back to read a *Wall Street Journal* article from three months ago; then teleport over to all my e-mail from that day; and then notice I'd forwarded the article to someone along with a couple of cool ideas, ideas I'd forgotten I'd ever written. It's like software for productive daydreaming.

Facetmap is based on a truth psychologists have long understood: We organize our memories by time and people. Those categories, Czerwinski says, are the pathways into the forest and the portholes from one memory to another.

"The way you remember things is associative," she says. "You think, 'There's all this stuff in my life that's related to Clive, or to Gordon.' Or you think, 'There's all this stuff that happened last fall.'" If you vaguely remember a book but can't recall the title or author, the first thing you're likely to begin with is the friend who told you about it; so you hunt through all the e-mails you got from him.

One of Czerwinski's colleagues at Microsoft, Susan Dumais, analyzed how people search for things on their computers and found that about one-quarter of the queries were for someone's name. She also explored ways to use "landmark" events to index our memories. "You'll think, 'Oh, I'm sure it happened right before the wedding, or just after Hurricane Katrina,'" Dumais says. She developed an experimental piece of software that embedded those landmarks into search tools so that you could, say, start with a major event and then see all the e-mail or Web pages you looked at that day. It drastically improved the ability to find things in the distant past, she says.

These sorts of tricks are already helping Bell find his memories. Gemmell has written a piece of software that works much like Facetmap. It's less graphical (it looks more like a regular Windows search) but just as powerful, as Gemmell illustrated with an example during my visit to his office: At one point last year, Bell had considered selling some property, so he surfed a bunch of real-estate Web pages and chatted with Connie, his broker. A few months later, he wanted to revisit those pages but couldn't figure out the right key words to bring them up. He had Connie's number, though, so he located the copy of his call to her. Then he checked the rest of his activity for that call, and presto: There were the Web sites, too.

"It works via how your memory works," Gemmell says. "It's like, 'I don't remember the specific words on that real-estate page.' My memory is just, 'It's the page I saw when I was talking to Connie.' We have to make 'search' more like the way we actually think."

INSTANT REPLAY

I came away from these demos eager for this stuff to come out of the lab and into the world. But the fact is that our everyday search tools are already moving slowly in this direction. In the past year, free "desktop search" programs by Google and Microsoft—which scour your hard drive— have begun incorporating sophisticated filters that let you work in similar ways: You can start by looking for a person, then find all the memos you've written to them, and then quickly zoom in on a day.

One day when I met Bell in San Francisco, I got a

chance to see his life through his eyes. He'd worn his Sense-Cam to work that day, and when he plays the images back rapidly, it's like watching a crude, stop-motion movie: buying coffee at Starbucks, grabbing a paper, entering his building, and finally dropping down at his desk.

"If you lose your keys, you can scroll back and figure out where you put 'em," he jokes. In fact, Bell seems like the sort of guy who might lose his keys a lot. He'd regularly get halfway through a sentence and then cut himself off and race along another tangent, only to have his hamster brain veer away halfway through that thought, too. *The guy's obviously crazy-smart, I thought. But no wonder he loves having a camera record the messy details of life.*

Yet here's the problem with the pictures: They pose an even bigger search dilemma, because computers can't "see" the contents of a photo. It's impossible for Bell to hunt for "pictures of my desk at work" or "that tall blond guy I met at the party"; at best, he can sort them by date or GPS coordinates. And while he has added keyword "tags" to many shots, it's time consuming and still not terribly accurate. Even he admits he rarely peruses any of his thousands of SenseCam pictures.

So are all those photos a waste of memory? Or can that kind of exhaustive visual record actually be worth something?

Alan Smeaton, a professor of computing at Dublin University, thinks it can. After hearing about Bell's project, Smeaton got Microsoft to lend him a few SenseCams and gave them to his students, who began wearing them all day long. They discovered an intriguing psychological effect: If, at the end of each workday, they spent a minute scrolling through the thousands of pictures the SenseCam had taken—a high-speed replay of their day—it had the effect of stimulating their short-term memory.

"You actually remember things you'd already forgotten," Smeaton says. "You'd see somebody you met in a corridor and had a two-minute conversation with that you'd completely forgotten about. And you'd go, 'Oh, I forgot to send an e-mail to that guy!' It's bizarre. It improves your recall by 100%."

In fact, "refresher" imagery is so powerful that it seems to help restore recall in people who have very little memory or none at all. Ken Wood, a computer scientist in Microsoft's research lab in Cambridge, gave a SenseCam to a UK woman who had lost her short-term memory due to encephalitis. She began wearing it to events she wanted to retain. After each of them, she reviewed the pictures several times over the course of the following two weeks. When the researchers quizzed her a month later, she still had "significant recall."

"She was over the moon," Wood marvels. "Totally thrilled."

SILICON CORTEX

Consider for a second how, precisely, we think. We use our memories all the time, of course, often by "active" remembering—scrolling through our minds to locate a tidbit. But much mental labor is passive. We think about something in the background, subconsciously letting a problem brew. Then we suddenly hit upon an interesting combination of things, a new way of thinking about a problem: the elusive, all-important epiphany.

What if our computers had their own intelligence and could do that background work for us? What if they could mine our memories for new ways of thinking? And what if they could prioritize the vast heaps of material in the backs of our minds, shaping the informational chaos that often leaves Bell so baffled? A memory system that could think on its own would unlock the lifelog's full potential.

Already Bell and Gemmell have played around with this effect using the SenseCam data by developing a screen saver that displays random snapshots from their personal archives. Bell finds it oddly mesmerizing: Pictures of long-ago birthdays or family trips will trigger waves of nostalgia, he says. But Czerwinski predicts that a similar screen saver could become a killer app in the office. When you're working on a project, the screen saver would cycle randomly through any documents, pictures, e-mails, or Web pages pertaining to your work—and you would see if the unpredictable combinations inspired fresh ways of understanding it. "You'd see some memo you wrote two years ago and think, 'Oh, right, that was a good idea. Why didn't I follow up on that?'" she says.

But the real goal is "to discover things that even you didn't know that you knew," says Bradley Rhodes, a computer scientist with Ricoh Innovations. In the lifelogging community, Rhodes is famous for creating the "Remembrance Agent," an experimental piece of software, as a PhD student at MIT. The Agent sits in the corner of your screen and pays attention to everything you type; every few seconds it checks inside your hard drive to see if it can find anything relevant. If it does, it alerts you in the corner of your screen by showing a line or two of the related document.

The connections the Agent discovers are surprising, often valuable. When Rhodes first started using it, he'd begin writing an e-mail to ask a colleague a question, but before he could even push "send," the Agent would reveal that a long-forgotten document on his hard drive already contained the answer. Other times, a colleague would e-mail him a question, and the Agent would remind Rhodes that he'd been asked that once before and had forgotten to reply.

"So, in mid-e-mail, I realize I have to switch gears and apologize and go, 'Sorry for not getting back to you.' It actually would change my behavior," he says. By actively reminding you of things from the past, "it keeps you from looking stupid." Now there's a killer app.

After using the Remembrance Agent for a while, Rhodes got a reputation for uncanny recall. "I'd have people e-mailing me saying, 'Hey, Brad, I know you've got this augmented brain. Can you answer this?' And sometimes I could." He imagines the Agent as particularly useful for lawyers and other paranoid execs drowning in paperwork. They could cram every sensitive e-mail and memo related to a corporate crisis into their computers and let the Agent monitor them on the periphery.

One of Bell's Microsoft allies is also investigating whether artificial intelligence could be used to find hidden patterns in memory. One day last summer, I visited Eric Horvitz, an expert in machine intelligence at the Research Labs, to see his "Lifebrowser." The Lifebrowser's goal is to automatically identify the most significant events in your life, so that when you scroll back through your history, it shows you only the most important highlights.

The software starts by having you "train" it by rating different things on your computer as significant or negligible. But then you turn Lifebrowser loose, and things get really interesting. It begins to observe your daily behavior and rank your documents and calendar events. Horvitz has given the software to members of his team at Microsoft, and Lifebrowser has already discovered several intriguing things about how we work. Unusual events—"atypia"— tend to be the ones that people most want to remember. Intuitively, this makes sense: Sudden, unexpected news, good or bad, is highly meaningful; a large cluster of pictures from the same day usually means you saw something odd or important—pluck one of them out and it's likely to be a rich memory. (In contrast, regular meetings are virtually never remarkable. "No one ever needs to remember what happened at the regular Monday staff meeting," Horvitz notes dryly.)

He pulls me over to his computer screen to see his own personal Lifebrowser. At first, it just looks like a regular calendar, with months stretching back years. But when he zooms in on October 2004, I can see that Lifebrowser has carefully picked only a few items to display on each day: A meeting at Defense Advanced Research Projects Agency (DARPA), the military-research agency. Pictures of Horvitz's family visit to Whidbey Island. An e-mail announcing a surprise visit from his old college friend.

"I would never have thought about this stuff myself," Horvitz says, scanning the calendar. "But as soon as I see it, I'm like, 'Oh, right—that was important!'" Lifebrowser, in essence, shapes the seemingly random flow of our lives and reminds us of what we ought to be focusing on.

For me, though, the real holy grail would be a system that doesn't merely remind me of what I already know— or once knew, even. It would be one that actually conceives of new ideas.

Few computer scientists are claiming to have figured that one out. But a hint of what may be coming has already arrived from Devon Technologies, a company that has created a tool called DEVONthink. It works like this: As you work on a project, you feed DEVONthink copies of any e-mails, memos, documents, PDF files, or Web pages you find interesting. The software performs a complex analysis of all that information, trying to find documents that are statistically similar to one another in meaning. This is much more sophisticated than a simple word search, though: DEVONthink can learn that the words *automobile repair* and *car fixing* are synonymous, for example, even though they use completely different terms. This allows the software to spot some astonishingly subtle connections.

Steven Johnson, a writer in New York, discovered that DEVONthink's smarts could literally inspire him to come up with profitable new ideas. He had been using the software for several years, putting all his research into it. While writing his latest book last year on the London cholera outbreak of 1854, he'd plug paragraphs of his draft into DEVONthink to see if anything else on his hard drive was related. One day, he inserted a passage on the London sewer system that used the word *sewage* a lot. DEVONthink unexpectedly unearthed a quote that discussed how vertebrates originally evolved their bones—by reusing the calcium waste generated by their metabolism.

It was a weird connection: comparing the waste systems of a city to those of an animal. But a lightbulb went off in Johnson's head, and he began dreaming up a new chapter for his book, contrasting the ways that cities and the human body find productive uses for their by-products.

"Now, strictly speaking, who is responsible for that initial idea?" he wondered, when he described the experience in an article. "Was it me or the software?" The idea, he argues, was a synthesis of two minds—his real brain and his virtual one. "Two very different kinds of intelligence playing off each other," he says, "one carbon-based, the other silicon."

In spring 2004, Gemmell lost a chunk of his memory. The Microsoft senior researcher had built his own personal MyLifeBits database, filling it, like Bell, with oodles of his e-mail, Web surfing, and pictures. But one day, Gemmell's hard drive crashed, and he hadn't backed up in four months.

When he got his MyLifeBits back up and running, the hole that had been punched in his memories was palpable, even painful. He'd be working at a project, vaguely remember some Web site or document that was important, and then begin drilling down to find it—only to discover that it was part of the missing period.

"It was like having my memories stolen," he says. He was amazed to realize his backup brain was no longer some novelty but a regular part of his psychological landscape. "I realized I count on this now. It's like I expect to drive cars and have flush toilets."

Machine memory is obviously giving us astonishing abilities, but can we deal with the change it'll wreak in our lives and work? We might become so reliant on artificial memory that we lose the habit of noticing things. "What is going to happen to us if we bypass something and we ought to have noticed it?" wonders Frank Nack, the lifelogging skeptic. Is it possible to forget how to remember? Perhaps so: Most of us have lost a cell phone only to realize that we can no longer recall the phone numbers of even our closest friends because the machine remembered them for us.

Whatever it all means, Bell will likely be the first person on the planet to find out. As I leave him at his San Francisco office, he offers me a parting gift, and fittingly enough, it's a memory device: one of the notebooks he buys at a nearby stationery store. It's beautiful, oddly old-fashioned. Yes, the man who swore off paper knows it would be much easier for MyLifeBits if he did all his writing electronically, on a digital tablet or whatnot. "But I can't help it," he says, "I just love these gorgeous Japanese-made notebooks." He did, after all, spend decades as a young engineer recording every idea on pads of his own. Some habits die hard.

The cost of a gigabyte of computer memory, over time.

- 1956: $10 million
- 1980: $233,000
- 1990: $7,700
- 2000: $13.30
- 2006: $1

Sources: ALTS LLC and *PC World*. Figures not adjusted for inflation.

Jeffrey M. O'Brien

You're *Soooooooooo* Predictable

Everything you buy online says a little bit about you. And if all those bits get put into one big trove of data about you and your tastes? Marketer's heaven.

"*If a girl says she likes* The Big Lebowski, *instantly, I think 'stoner.'*" *That's Matthew Kuhlke speaking. We're sitting at a table full of grilled meat and jug sodas on a late-summer weeknight in midtown Atlanta. "She hangs out with a bunch of guys. She dates them a little bit, but she really just likes the attention."*

Also at the table: Kuhlke's business partner, Adam Geitgey. Kuhlke and Geitgey grew up in nearby Augusta and finish each other's thoughts the way lifelong friends often do. The pair of unattached, hormonal twenty-somethings also have a knack for turning conversations toward dating. "If all a girl likes are romantic comedies," Geitgey adds, "I'd be worried that she's going to be codependent, emotionally needy."

I've invited Kuhlke and Geitgey to the restaurant of their choosing to talk about movies and personality. They opted for the Vortex, a bar and grill with a skull for a logo and a rash of thrift-store salvage covering every last bit of wall space. Chairs dangle from the ceiling; a skeleton sits atop a motorcycle near an aluminum camel. As the waitstaff buzzes around in hipster flair—piercings, tattoos, dog collars—the former Georgia Tech roommates explain the underlying premise of their company, What to Rent.

The site administers a personality test to visitors and recommends DVDs based on the findings. To demonstrate the connection between movies and personality, the budding entrepreneurs have been describing prototypical fans of everything from Raging Bull *to* Finding Nemo. *Now they'll attempt something more difficult: They'll each pick someone in the restaurant and, without ever talking to them, divine their single favorite movie.*

Once they make their picks, my plan is to corner the unsuspecting strangers and see how perceptive these film nerds really are.

Recommender systems like Whattorent.com are sprouting on the Web like mushrooms after a hard rain. Dozens of companies have unveiled recommenders recently to introduce consumers to Web sites, TV shows, other people— whatever they can think of. The idea isn't new, of course. In a time of impersonal big-box stores and self-checkout stations, independent shopkeepers compete by doing this sort of thing every day. What does the fact that you drive a Prius and buy organic baby food say about you? At the farmer's market, it says, "Grass-fed ribeye, heirloom tomatoes, and line-caught salmon." In the local appliance store, you're looking for a low-energy, front-loading washer with matching dryer. For those shopkeepers, it's more than a parlor trick. It's good business.

We don't just buy products; we bond with them. We have relationships with our things. DVD collections, iTunes playlists, cars, cell phones: Each is an extension of who we are (or want to be). We put ourselves on display through our purchases, wearing our personalities on our sleeves, literally and figuratively, for the world to see. And if you don't subscribe to this sort of materialism—if you don't define your-self by the clothes on your back or the neighborhood you live in—well, that's just another brand of expression. In the real world, we use apparent information, coupled with context, experience, and stereotypes, to size each other up. This sort of intuition is useful and often accurate, but it's also fallible. Online, the picture becomes clearer. Consumers now routinely rank experiences on the Web—four stars on IMDb for *The Departed,* three stars on Epinions for a Roomba vacuum, a positive eBay rating, a Flickr tag. Each time you leave such a mark, you help the rest of us make sense of all the look-alike, sound-alike stuff on the Web.

You also leave a trail. For the company that can decipher all that information, the opportunity is staggering. That company will know you better than the shopkeeper knows you, better than the credit bureau or, arguably, even your spouse. It will pinpoint your tastes and determine the likelihood that you'll buy any given product. In effect, it will have constructed the algorithm that is you.

There's a sense among the players in the recommendation business—from newcomers like MyStrands and StumbleUpon to titans like Yahoo and Sun—that now is the time to perfect such an algorithm. The Web, they say, is leaving the era of search and entering one of discovery. What's the difference? Search is what you do when you're looking for something. Discovery is when something wonderful that you didn't know existed, or didn't know how to ask for, finds you. When it comes to search, there's a clear winner—a $145 billion company called Google. But there is no go-to discovery engine—yet.

Building a personalized discovery mechanism will mean tapping into all the manners of expression, categorization, and opinions that exist on the Web today. It's no easy feat, but if a company can pull it off and make the formula portable so it works on your mobile phone—well, such a tool could change not just marketing but all of commerce. "The effect of recommender systems will be one of the most important changes in the next decade," says University of Minnesota computer science professor John Riedl, who built one of the first recommendation engines in the mid-1990s. "The social web is going to be driven by these systems."

Amazon realized early on how powerful a recommender system could be and to this day remains the prime example. The company uses a series of collaborative filtering algorithms to compare your purchasing patterns with everyone else's and thus narrow a vast inventory to just the stuff it predicts you'll buy. "Personalized recommendations," says Brent Smith, Amazon's director of personalization, "are at the heart of why online shopping offers so much promise."

So far, the company has struggled to deliver on that promise. Its system favors popular, obvious items and tends to come off less like a trusted shopkeeper than a pushy salesman: If you liked the novel *The Corrections,* you'll see suggestions to buy *The Discomfort Zone* and everything else Jonathan Franzen has written. If you bought a gift for a baby shower, you're bound to get a stream of recommendations for cheap plastic toys and birthing blankets. The new generation of recommenders will do better. Some employ filters that factor in more variables. Others analyze the contents of what they're recommending to grasp why you like something. A third category, hybrid recommenders, combines both strategies.

To tap into some of the brainpower gathering in the recommender space, Netflix recently established the Netflix Prize, offering a $1 million bounty to anyone who can improve by 10 percent the efficacy of the company's recommender system. A few days before the contest was officially announced, VP of recommendation systems Jim Bennett expressed doubt about whether anyone would reach the goal ahead of the ten-year deadline but said it'd be well worth the $1 million if someone did. Five weeks later, 37 registrants had already posted improvements to the Netflix system. Two contestants were just short of halfway to the goal.

Back at the Vortex, Kuhlke and Geitgey discuss how their parlor trick works. They'll draw on their knowledge of cinema and their experience categorizing hundreds of films—by star power, plot complexity, etc.—at What to Rent. "When you watch a movie, you interact with it like you're interacting with another person," Kuhlke says. "You're forming a relationship."

Geitgey goes first. He scans the crowd and homes in on a target, a guy delivering meals and clearing dishes. On their site, Kuhlke and Geitgey use a series of odd questions to determine a user's personality. "How much money would it take for you to wear a neon-green fanny pack for the rest of your life?" Or: "What do you like better, reading a book or watching TV?" Here, it's all about observation. This is what Geitgey sees: tattered jeans, a steel bracelet, a few tattoos. The target may be gathering dishes but seems too authoritative to be a busboy. He appears to be in his late 20s and is working, Geitgey surmises, "in a youth-trendy restaurant in the part of the city where people that age who don't have real jobs hang out."

Geitgey makes a few leaps. "Those are the kind of guys who barely made it through high school because they couldn't focus but spend most of their time reading light philosophy books by singers-turned-writers like Nick Cave," he says, insisting that the target is trying to square budding intellectualism with his physical image.

So. Which movie? "He would be interested in things that have an underlying philosophy but are also physically intense," Geitgey concludes. "Starship Troopers fits that exactly—a bit of mild antiestablishment philosophy with some bad-ass bug-killing."

In a few short years, Pandora has become the most efficient new-music discovery mechanism in history. That's not saying much, really. Consider the alternatives: scouring magazines for reviews, flipping through albums in the record store, listening to radio stations all play the same songs. At Pandora.com, you type in the name of a band or song and immediately begin hearing similar tunes that the site's recommender system—aka the Music Genome Project—has determined you'll enjoy. By rating songs and artists, you can refine the suggestions, allowing Pandora to create a truly personalized station.

Unlike collaborative filtering engines, Pandora understands each song in its database. Forty-five analysts, many with music degrees, rank 15,000 songs a month on 400 characteristics to gain a detailed grasp of each. A former musician and film composer, founder Tim Westergren came up with the idea for Pandora while scoring movies for directors. "They weren't musicians, so no one was saying, 'I like minor harmonies and woodwinds,'" he says, leaning back in his chair at Pandora's Oakland headquarters. "So it was my job to figure out their musical taste. I developed a genome in my head and would say, 'Okay, you like this song; do you like this one?'"

Four million people now use Pandora the same way those directors used Westergren. Let's say I type in Bloc Party, a pop/punk band that's part Franz Ferdinand, part Sonic Youth. The next tune I hear might be from We Are Scientists, a Brooklyn indie band with, Pandora determines, similar "electric-rock instrumentation, subtle use of vocal harmony, and minor-key tonality." Next I get a catchy song from a pop-punk/emo band called Fire When Ready, which stands at No. 225,301 on Amazon's music sales list. It's safe to say that few consumers are searching for this band. But on Pandora, it's only a few degrees of separation from Bloc Party.

To draw a line from Bloc Party to Fire When Ready, the Music Genome Project combs through hundreds of thousands of songs and millions of pieces of user feedback. It's an impressive technological accomplishment but not nearly as impressive as the implications. If Pandora can nail me as a fan of a band that few people have ever heard of, and my musical tastes are an intimate expression of who I am, then Pandora could introduce me to a lot more than music. Take it from Jason Rentfrow. "If you know I like Ahmad Jamal," Rentfrow says, referring to the 64-year-old jazz musician who played piano for Miles Davis, "that'll stimulate other information that you can infer about me."

So Rentfrow likes sophisticated jazz. What sort of picture does that paint? If you think he's intelligent, articulate, a bit geeky and soft-spoken, and wears glasses, well, you'd be relying on stereotypes. And you'd be right. Of course no person is one dimensional. Rentfrow also likes the rap group A Tribe Called Quest. With that information, you would probably redraw his caricature. Now maybe he seems younger and more open-minded. Revealing a list of his 50 favorite artists would flesh him out even more.

Rentfrow is a 30-year-old psychology professor at the University of Cambridge (Britain). To study the links between musical taste and personality, he and University of Texas psychology professor Sam Gosling administered personality tests to 74 students and instructed each to submit a list of 10 favorite songs, which were then played for another set of volunteers. After listening to the songs, the second group ranked each person on 28 characteristics. Rentfrow then compared those results with the earlier personality tests. Music turned out to be a poor predictor of emotional stability, courage, and ambition but accurate on extroversion, agreeableness, conscientiousness, openness, imagination, and even intellect. An ongoing study conducted by Rentfrow at www.outofservice.com, where 90,000 people have taken a music/personality quiz, pushes the point even further, tying taste to political leanings, demographics, lifestyle, favorite authors—movies even.

As recently as a decade ago, when musical preference was the province of a stack of CDs in the living room, such research may have been a dead end. But music recommenders can now draw lines from musical tastes to all sorts of things. Rentfrow and Gosling explored the connection between music and personality out of scientific curiosity. But the connection could have broad applications in business. Good-bye, context-based advertising. Hello, personality-based advertising.

Geitgey's made his choice: "Starship Troopers." Now it's Kuhlke's turn. He's spotted a waitress. She's in her late teens or early 20s, black hair in a bob, very cute. She looks at the floor as she walks and avoids eye contact with customers. "She's unhappy," Kuhlke surmises. "She's working around all these jerks who just want to have sex with her."

Geitgey chimes in, suggesting she wasn't popular in high school and is shaking off her past by working in a cool place. Kuhlke cuts him off. "This place is the Applebee's of cool!" he insists. "If it was really cool, it wouldn't be a chain, and there wouldn't be a paid parking lot across the street."

So what? "So she'd pick the wrong movie," he continues. "She'd like an old-school romantic comedy, but she'd pick Breakfast at Tiffany's, *a totally crappy movie. She's cool enough to know she should pick Audrey Hepburn, but not cool enough to pick the right movie,* Roman Holiday."

The waitress is obviously attractive, maybe the prettiest woman in the restaurant. And she does seem bothered by—or at least indifferent to—her surroundings. But that hardly makes her prone to ill-informed choices. Kuhlke reconsiders. Maybe she would *like* Roman Holiday. *He clearly wants her to say* 'Roman Holiday.' *Flustered, he jumps off the romantic comedy train altogether.*

She's disaffected, he insists, throwing his arms in the air. "Girl, Interrupted."

Just back from a visit to his childhood home in Kiev, Max Levchin is in his South of Market office holding up a propaganda handbill that he picked up in Moscow. He wants to mount it on a wall at his new startup, Slide. The poster features a railroad worker staring out from a caboose and a headline that Levchin translates aloud. "Literally, it says 'Night—not a hindrance to work!'" he says, mocking a sinister laugh, "though the implication is clearly, 'Keep working, bitches!'"

Levchin's team doesn't really need the reminder. The 31-year-old CEO drinks eight espressos a day and possesses a work ethic that's famous around Silicon Valley. He often arrives in the office by six and is still there 12 or even 18 hours later. After cofounding the online payment service PayPal in 1998, he battled fraudsters before taking the company public in 2002 and selling it several months later to eBay for $1.5 billion. "The crux of PayPal is a risk-management company," he explains. "We perfected the ability to say, 'The odds of you stealing money are X.'"

Levchin cross-referenced traffic patterns, auction histories, user interactions, geography, and a thousand other factors to root out fraud. "I would obsess on multivariate analysis. I'd have printouts and printouts graphing the relationship between any two variables," he says. "For a long time we didn't even patent this stuff because it was so secret."

These days, Levchin has the typical angst of a second-time entrepreneur. He's obsessed with proving that his first success wasn't a fluke, and he wants to exploit the biggest opportunity he can find. He's using what he learned at Pay-Pal, not to root out fraud but to create the best recommender system he can imagine, one that will cover the entire Web, pulling content of all kinds—music, movies, gadgets, blogs, news stories, cars, one-night stands, you name it—filtering it according to individual preference and delivering it to the desktop. Instead of quantifying the odds of your stealing money, he's building a "machine that knows more about you than you know about yourself."

If Slide is at all familiar, it's as a knockoff of Flickr, the photo-sharing site. Users upload photos, which are displayed on a running ticker or Slide Show, and subscribe to one another's feeds. But photos are just a way to get Slide users communicating, establishing relationships, Levchin explains. The site is beginning to introduce new content into Slide Shows. It culls news feeds from around the Web and gathers real-time information from, say, eBay auctions or Match.com profiles. It drops all of this information onto user desktops and then watches to see how they react.

Suppose, for example, there's a user named Yankee-Dave who sees a Treo 750 scroll by in his Slide Show. He gives it a thumbs-up and forwards it to his buddy, we'll call him Smooth-P. Slide learns from this that both YankeeDave and Smooth-P have an interest in a smartphone and begins delivering competing prices. If YankeeDave buys the item, Slide displays headlines on Treo tips or photos of a leather case. If Smooth-P gives a thumbs-down, Slide gains another valuable piece of data. (Maybe Smooth-P is a BlackBerry guy.) Slide has also established a relationship between Yan-keeDave and Smooth-P and can begin comparing their ratings, traffic patterns, clicks, and networks.

Based on all that information, Slide gains an understanding of people who share a taste for Treos, TAG Heuer watches, and BMWs. Next, those users might see a Dyson vacuum, a pair of Forzieri wingtips, or a single woman with a six-figure income living within a 10-mile radius. In fact, that's where Levchin thinks the first real opportunity lies—hooking up users with like-minded people. "I started out with this idea of finding shoes for my girlfriend and hotties on HotorNot for me," Levchin says with a wry smile. "It's easy to shift from recommending shoes to humans."

If this all sounds vaguely creepy, Levchin is careful to say he's rolling out features slowly and will only go as far as his users will allow. But he sees what many others claim to see: Most consumers seem perfectly willing to trade preference data for insight. "What's fueling this is the desire for self-expression," he says.

As Levchin and his charges set out to build a new type of search engine—a new Google—one question becomes obvious: Where's the old Google? VP of engineering Udi Manber refuses to comment on whether there's a recommender system in the pipeline. But it's a safe bet something is coming. Google's director of research, Peter Norvig, is an adviser to CleverSet, a recommender company in Seattle. Steve Johnson, CEO of Boston-based recommender system ChoiceStream, which provides the collaborative filtering engine behind AOL.com, Blockbuster.com, iTunes, and Directv.com, says he's been talking to Google about building a system for YouTube. "Google needs to go to a preference-based search paradigm, and I believe they're moving in this direction," Johnson says.

Kuhlke and Geitgey never had Google-sized ambitions for
Whattorent.com. The cofounders, who recently took day jobs as
software engineers on opposite ends of the country, didn't expect
their site to be a huge moneymaker. (And so far, it hasn't been.
"We've made exactly zero dollars," Kuhlke says.) They were a
couple of undergrads trying to save their pals from the movie
geek's ultimate nightmare scenario—walking around Block-
buster with no idea what to rent. They accomplished that much.
But they may have also helped pave the way for a whole new
way for marketers to get inside your head.

We've paid the bill, and it's the moment of truth. We
approach the guy with tattoos, who, as it turns out, is the restau-
rant manager. We tell him what we've been up to and ask him to
tell us his favorite movie without thinking.

"Brianna Loves Jenna," he says with half a smile and a lift
of the eyebrows. Everyone laughs. New rule: no porn. He crosses
his arms, pauses for a moment, and exclaims, "Starship Troop-
ers!"

High-fives all around. Geitgey nailed it.

Now the waitress. We haven't seen her talk to anyone since
we walked in. Kuhlke and Geitgey hang back cautiously as I
approach her. I let her in on our little parlor trick, asking her to
quickly tell me her favorite movie. And for the first time all
night, the woman who will be remembered as "Audrey" lifts her
cheeks to reveal a smile that comes both from a set full of won-
derful white teeth and a pair of impossibly green eyes. She says
two words.

"Roman Holiday."

Joshua Davis

Say Hello to Stanley

*Stanford's souped-up Volkswagen blasted through
the Mojave Desert, blew away the competition,
and won the DARPA $2 million Grand Chal-
lenge. Buckle up, human—the driverless car of the
future is gaining on you.*

Sebastian Thrun is sitting in the passenger seat of a 2004
Volkswagen Touareg that's trying to kill him.

The car hurtles down a rutted dirt road at 35 miles per
hour somewhere in the Mojave Desert, bucking and swerv-
ing, kicking up a cloud of dust. Thrun, the youngest person
ever to head Stanford's famed artificial intelligence laboratory, clings to an armrest. Mike Montemerlo, a speed-coding
computer programmer and postdoc, is wedged in the backseat amid a tangle of wires and cables.

No one is driving. Or more precisely, the Touareg is trying to drive itself. But despite 635 pounds of gear—roof-
mounted radar, laser range finders, video cameras, a seven-
processor shock-mounted computer—the car is doing a
lousy job. Thrun tightens his grip on the armrest. He's built
plenty of robots, but he's never entrusted his life to one of his
creations. He's scared, confused, and above all furious that
his algorithms are failing.

Suddenly the steering wheel spins itself hard to the left, and the car speeds toward a ditch. David Stavens, a programmer who is stationed in the driver's seat in case of emergency, grabs the wheel and fights the pull of the robotic autopilot, which is insisting on a plunge into the gulley. Stavens slams his foot down on the computer-controlled brake. Thrun hits the big red button on the console that disables the vehicle's navigation computers. The SUV skids to a halt. "Hey, that was exciting," Thrun says, trying to sound upbeat.

It wasn't supposed to be this way. In 2003, the Defense Advanced Research Projects Agency (DARPA) offered $1 million to anyone who could build a self-driving vehicle capable of navigating 300 miles of desert. Dubbed the Grand Challenge, the robot-vehicle race was hyped for months. It was going to be as important as the 1997 Kasparov–Deep Blue chess match. But on race day in March 2004, the cars performed like frightened animals. One veered off the road to avoid a shadow. The largest vehicle—a 15-ton truck—mistook small bushes for enormous boulders and slowly backed away. The favorite was a Carnegie Mellon University (CMU) team that, fueled by multimillion-dollar military grants, had been working on unmanned vehicles for two decades. Its car went 7.4 miles, hit a berm, and caught fire. Not a single car finished.

Back at Stanford, Thrun logged on to check the progress of the race and couldn't believe what he was seeing. It was a humiliation for the entire field of robotics—a field Thrun was now at the center of. Only a year before, he'd been named head of Stanford's AI program. In the quiet halls of the university's Gates Computer Science Building, the suntanned 36-year-old German was a whirlwind of excitement, ideas, and brightly colored shirts. He was determined to show what intelligent machines could contribute to society. And though he had never considered building a self-driving car before, the sorry results of the first Grand Challenge inspired him to give it a try.

He assembled a first-rate team of researchers, attracted the attention of Volkswagen's Palo Alto R&D team, and charged ahead. But here in the desert, he's facing the reality that the Touareg—dubbed Stanley, a nod to Stanford—is totally inadequate. With only three months to go before the second Grand Challenge, he realizes that some basic problems remain unsolved.

Thrun gets out to kick the dirt on the side of the road and think. While the car idles, he squints at the uneven terrain ahead. This was his chance to lead the way toward his vision of the new vehicular order. But for now, all he sees is mountains, sagebrush, and sky.

It started with a black-and-white video game in 1979. Thrun, then 12, was spending most of his free time at a local pub in Hannover, Germany. The place had one of the first coin-operated video games in town, and 20 pfennig bought him three lives driving at high speed through a stark landscape of oil slicks and oncoming cars. It was thrilling—and much too expensive. For weeks, Thrun scrutinized the graphics and then decided that he could re-create the game on his Northstar Horizon, a primitive home computer that his father, a chemical engineer, had bought for him. He shut himself in his room and devoted his young life to coding the Northstar. It ran at 4 MHz and had only 16 Kbytes of RAM, but somehow he coaxed a driving game out of the machine.

Though he didn't study or do much homework over the next seven years, Thrun ended up graduating near the top of his high school class. He wasn't sure what was next. He figured he'd think about it during his mandatory two-year stint in the German army. But on June 15, 1986—the last day to apply for university admissions—military authorities told him he wouldn't be needed that year. Two hours later, he arrived at the centralized admission headquarters in Dortmund with only 20 minutes to file his application. The woman behind the counter asked him what he wanted to study—in Germany, students declare a major before arriving on campus. He looked down the list of options: law, medicine, engineering, and computer science. Though he didn't know much about computer science, he had fond memories of programming his Northstar. "Why not?" he thought and decided his future by checking the box next to computer science.

Within five years, he was a rising star in the field. After posting perfect scores on his final undergraduate exams, he went on to graduate school at the University of Bonn, where he wrote a paper showing for the first time how a robotic cart, in motion, could balance a pole. It revealed an instinct for creating robots that taught themselves. He went on to code a bot that mapped obstacles in a nursing home and then alerted its elderly user to dangers. He programmed robots that slithered into abandoned mines and came back hours later with detailed maps of the interior. Roboticists in the United States began to take note. Carnegie Mellon offered the 31-year-old a faculty position and then gave him an endowed chair. But he still hadn't found a research area to focus all his energy and skills on.

While Thrun was settling in at CMU, the hot topic in robotics was self-driving cars. The field was led by Ernst Dickmanns, a professor of aerospace technology at the University of the Bundeswehr. He liked to point out that planes had been flying themselves since the 1970s. The public was clearly willing to accept being flown by autopilot, but nobody had tried the same on the ground. Dickmanns decided to do something about that.

With help from the German military and Daimler-Benz, he spent seven years retrofitting a boxy Mercedes van, equipping it with video cameras and a bunch of early Intel processors. On a Daimler-Benz test track in December 1986, the driverless van accelerated to 20 miles per hour and, using data supplied by the videocams, successfully stayed on a curving road. Though generally forgotten, this was the Kitty Hawk moment of autonomous driving.

It sparked a 10-year international dash to develop self-driving cars that could navigate city streets and freeways. In the United States, engineers at Carnegie Mellon led the charge with funding from the army. On both sides of the Atlantic, the approach involved a data-intensive classification approach, a so-called rule-based system. The researchers assembled a list of easily identifiable objects (solid white lines, dotted white lines, trees, boulders) and told the car what to do when it encountered them. Before long, though, two main problems emerged. First, processing power was anemic, so the vehicle's computer quickly became overwhelmed when confronted with too much data (a boulder beside a tree, for instance). The car would slow to a crawl while trying to apply all the rules. Second, the team couldn't code for every combination of conditions. The real world of streets, intersections, alleys, and highways was too complex.

In 1991, a CMU computer science PhD student named Dean Pomerleau had a critical insight. The best way to teach cars to drive, he suspected, was to have them learn from the experts: humans. He got behind the wheel of CMU's sensor-covered, self-driving Humvee, flipped on all the computers, and ran a program that tracked his reactions as he sped down a freeway in Pittsburgh. In minutes, the computers had developed algorithms that codified Pomerleau's driving decisions. He then let the Humvee take over. It calmly maneuvered itself on Pittsburgh's interstates at 55 miles per hour.

Everything worked perfectly until Pomerleau got to a bridge. The Humvee swerved dangerously, and he was forced to grab the wheel. It took him weeks of analyzing the data to figure out what had gone wrong: When he was "teaching" the car to drive, he had been on roads with grass alongside them. The computer had determined that this was among the most important factors in staying on the road: Keep the grass at a certain distance and all will be well. When the grass suddenly disappeared, the computer panicked.

It was a fundamental problem. In the mid-'90s, microchips weren't fast enough to process all the potential options, especially not at 55 miles per hour. In 1996, Dickmanns proclaimed that real-world autonomous driving could "only be realized with the increase in computer performance. . . . With Moore's law still valid, this means a time period of more than one decade." He was right, and everyone knew it. Research funding dried up, programs shut down, and autonomous driving receded back to the future.

Eight years later, when DARPA held its first Grand Challenge, processors had in fact become 25 times faster, outpacing Moore's law. Highly accurate GPS instruments had also become widely available. Laser sensors were more reliable and less expensive. Most of the conditions Dickmanns had said were necessary had been met or exceeded. More than a hundred contestants signed up, including a resurgent CMU squad. DARPA officials couldn't hide their excitement. The breakthrough moment in autonomous driving was, they thought, at hand. In truth, some of the field's biggest challenges had yet to be overcome.

Once Thrun decided to take a crack at the second Grand Challenge, he found himself consumed by the project. It was as though he were 12 again, shut up in his room, coding driving games. But this time a Northstar home computer wasn't going to cut it. He needed serious hardware and a sturdy vehicle.

That's when he got a call from Cedric Dupont, a scientist at Volkswagen's Electronics Research Laboratory, just a few miles from the Stanford campus. The Volkswagen researchers wanted in on the Grand Challenge. They'd heard that Thrun was planning to enter the event, and they offered him three Touaregs—one to race, another as a backup, and a third for spare parts. The VW lab would outfit them with steering, acceleration, and braking control systems custom-built to link to Thrun's computers. Thrun had his vehicle, and Volkswagen executives had a chance to be part of automotive history.

It was history, however, that Red Whittaker planned on writing himself. Whittaker, the imposing, bald, bombastic chief of CMU's eponymously named Red Team, had been working on self-driving vehicles since the '80s. Whittaker's approach to problem solving was to use as much technological and automotive firepower as possible. Until now, the firepower hadn't been enough. This time, he would make sure that it was.

First, he entered two vehicles in the race: a 1986 Humvee and a 1999 Hummer. Both were chosen for their ruggedness. Whittaker also stabilized the sensors on the trucks with gyroscopes to ensure more reliable data. Then he sent three men in a laser-studded, ground-scanning truck into the desert for 28 days. Their mission: create a digital map of the race area's topography. The team logged 2,000 miles and built a detailed model of the desolate sagebrush expanses of the Mojave.

That was only the beginning. The Red Team purchased high-resolution satellite imagery of the desert, and, when

DARPA revealed the course on race day, Whittaker had 12 analysts in a tent beside the start line scrutinize the terrain. The analysts identified boulders, fence posts, and ditches so that the two vehicles would not have to wonder whether a fence was a fence. Humans would have already coded it into the map.

The CMU team also used Pomerleau's approach. They drove their Humvees through as many different types of desert terrain as they could find in an attempt to teach the vehicles how to handle varied environments. Both SUVs boasted seven Intel M processors and 40 Gbytes of flash memory—enough to store a world road atlas. CMU had a budget of $3 million. Given enough time, manpower, and access to the course, the CMU team could prepare their vehicles for any environment and drive safely through it.

It didn't cut it. Despite that 28-day, 2,000-mile sojourn in the desert, CMU's premapping operation overlapped with only 2 percent of the actual racecourse. The vehicles had to rely on their desert training sessions. But even those didn't fully deliver. A robot might, for example, learn what a tumbleweed looks like at 10 a.m., but with the movement of the sun and changing shadows, it might mistake that same tumbleweed for a boulder later in the day.

Thrun faced these same problems. Small bumps would rattle the Touareg's sensors, causing the onboard computer to swerve away from an imagined boulder. It couldn't distinguish between sensor error, new terrain, its own shadow, and the actual state of the road. The robot just wasn't smart enough.

And then, as Thrun sat on the side of that rutted dirt road, an idea came to him. Maybe the problem was a lot simpler than everyone had been making it out to be. To date, cars had not critically assessed the data their sensors gathered. Researchers had instead devoted themselves to improving the quality of that data, either by stabilizing cameras, lasers, and radar with gyroscopes or by improving the software that interpreted the sensor data. Thrun realized that if cars were going to get smarter, they needed to appreciate how incomplete and ambiguous perception can be. They needed the algorithmic equivalent of self-awareness.

Together with Montemerlo, his lead programmer, Thrun set about recoding Stanley's brain. They asked the computer to assess each pixel of data generated by the sensors and then assign it an accuracy value based on how a human drove the car through the desert. Rather than logging the identifying characteristics of the terrain, the computer was told to observe how its interpretation of the road either conformed to or varied from the way a human drove. The robot began to discard information it had previously accepted—it realized, for instance, that the bouncing of its sensors was just turbulence and did not indicate the sudden appearance of a boulder. It started to ignore shadows and accelerated along roads it had once perceived as being crisscrossed with ditches. Stanley began to drive like a human.

Thrun decided to take the car's newfound understanding of the world a step further. Stanley was equipped with two main types of sensors: laser range finders and video cameras. The lasers were good at sensing ground within 30 meters of the car, but beyond that the data quality deteriorated. The video camera was good at looking farther away but was less accurate in the foreground. Maybe, Thrun thought, the laser's findings could inform how the computer interpreted the faraway video. If the laser identified drivable road, it could ask the video to search for similar patterns ahead. In other words, the computer could teach itself.

It worked. Stanley's vision extended far down the road now, allowing it to steer confidently at speeds of up to 45 miles per hour on dirt roads in the desert. And because of its ability to question its own data, the accuracy of Stanley's perception improved by four orders of magnitude. Before the recoding, Stanley incorrectly identified objects 12 percent of the time. After the recoding, the error rate dropped to 1 in 50,000.

It's half past 6 in the morning on October 8, 2005, outside of Primm, Nevada. Twenty-three vehicles are here for the second Grand Challenge. Festooned with corporate logos, lasers, radars, GPS transponders, and video cameras, they're parked on the edge of the gray-brown desert and ready to roll. The early morning light clashes with the garish glow of the nearby Buffalo Bill's Resort and Casino.

Red Whittaker is beaming. His 12 terrain analysts have completed their two-hour premapping of the route, and the data has been uploaded to the two CMU vehicles via a USB flash drive. The stakes are high this year: DARPA has doubled the prize money to $2 million, and Whittaker is ready to win it and erase the memory of the 2004 debacle. Last night, he pointed out to the press that Thrun had been a junior faculty member in Whittaker's robotics lab at CMU. "My DNA is all over this race," he boasted. Thrun won't be baited by Whittaker's grandstanding. He focuses on trying to calm his own frayed nerves.

The race begins quietly: One by one, the vehicles drive off into the hills. A few hours later, the critical moment is captured in grainy footage. CMU's H1 is in the middle of a dusty white desert expanse. The camera slowly approaches—the image is pixelated and overexposed. It's the view from Stanley's rooftop camera. For the past 100 miles, the Touareg has been tailgating the H1, and now it pulls close. Its lasers scan the exterior of its competitor, revealing a ghostly green outline of side panels and a giant, sensor-stabilizing gyroscope. And then the VW rotates its steering wheel and passes.

DARPA has imposed speed limits of 5 to 25 miles per hour, depending on conditions. Stanley wants to go faster. Its lasers are constantly teaching its video cameras how to identify drivable terrain, and it knows that it could accelerate more. For the rest of the race, Stanley pushes up against the speed limits as it navigates through open desert and curving mountain roads. After six hours of driving, it exits the final mountain pass ahead of every other team. When Stanley crosses the finish line, Thrun catches his first sight of an undiscovered country, a place where robots do all the driving.

The 128-mile race is a success. Four other vehicles, including both of CMU's entries, complete the course behind Stanley. The message is clear: Autonomous vehicles have arrived, and Stanley is their prophet. "This is a watershed moment—much more so than Deep Blue versus Kasparov," says Justin Rattner, Intel's R&D director. "Deep Blue was just processing power. It didn't think. Stanley thinks. We've moved away from rule-based thinking in artificial intelligence. The new paradigm is based on probabilities. It's based on statistical analysis of patterns. It is a better reflection of how our minds work."

The breakthrough comes just as carmakers are embracing a host of self-driving technologies, many of them barely recognizable as robotic. Take, for example, a new feature known as adaptive cruise control, which allows the driver to select the distance to be maintained between the vehicle and the car in front of it. On the Toyota Sienna minivan, this is simply another button on the steering wheel. What that button represents, however, is a laser that surveys the distance to the vehicle ahead of it. The minivan's computer interprets the data and then controls the acceleration and braking to keep the distance constant. The computer has, in essence, taken over part of the driving.

But even as vehicles are being produced with sensors that perceive the world, they have, until now, lacked the intelligence to comprehensively interpret what they see. Thanks to Thrun, that problem is being solved. Computers are nearly ready to take the wheel. But are humans ready to let them?

Jay Gowdy doesn't think so. A highly regarded roboticist, he has worked for nearly two decades to build self-driving cars, first with CMU and, more recently, with SAIC, a Fortune 500 defense contractor. He notes that in the United States, about 43,000 people die in traffic accidents every year. Robot-driven cars would radically reduce the number of fatalities, he says, but there would still be accidents, and those deaths would be attributable to computer error. "The perception is that in the majority of accidents today, those who die are drunk, lazy, or stupid and bring it on themselves," Gowdy says. "If computers take over the driving, any deaths are likely to be perceived as the loss of people who did nothing wrong."

The resulting liability issues are a major hurdle. If a robotically driven car gets in an accident, who is to blame? If a software bug causes a car to swerve off the road, should the programmer be sued, or the manufacturer? Or is the accident victim at fault for accepting the driving decisions of the onboard computer? Would Ford or GM be to blame for selling a "faulty" product, even if, in the larger view, that product reduced traffic deaths by tens of thousands?

This morass of liability questions would need to be addressed before robot cars could be practical. And even then, Americans would have to be willing to give up control of the steering wheel.

Which is not something they're likely to do, even if it means saving 40,000 lives a year. So the challenge for carmakers will be to develop interfaces that make people feel like they're in control even when the car is really doing most of the thinking. In other words, that small adaptive cruise control button in Toyota's minivan is a Trojan horse.

"OK, we're two of two, two of two, and one of one, no U-turn, speed advisory 25, large divider, POI gas station on left."

Michael Loconte and Bill Wong are creeping through a quiet suburb just north of San Jose, California. They are driving a white Ford Taurus with a six-inch antenna on the roof. Loconte wears a headset and mumbles coded descriptions of the surroundings into the microphone—two of two means that he's in the right lane on a street with two lanes, and POI" means point of interest. Wong scribbles with a digital pen, making landmark and street address notations on a scrolling map. "People think we're with the CIA," Loconte says. "I know it kind of looks like that."

But they aren't spies. They're field analysts working for the GPS mapping company Navteq, and they're laying the foundation for the future of driving. On this Friday afternoon, they're doing a huge commercial extension of CMU's ditch-and-fence mapping operation. Navteq has 500 such analysts driving U.S. neighborhoods, mapping them foot by foot. Though Thrun has proven that extensive mapping isn't needed to get from A to B, maps are critical when it comes to communicating with robotic vehicles. As automotive engineers build cars with increasing autonomy, the human interface with the vehicle will migrate from the steering wheel to the map. Instead of turning a wheel, drivers will make decisions by touching destinations on an interactive display.

"We want to move up the food chain," says Bob Denaro, Navteq's VP of business development. The company sees itself moving beyond the help-me-I'm-lost gizmo business and into the center of the new driving experience. That's not to say that the steering wheel will disappear; it will just be gradually de-emphasized. We will continue to sit in the driver's seat and have the option of intervening if we choose. As Denaro notes, "A person's role in the car is changing. People will become more planners than drivers."

And why not—since the car is going to be a better driver than a human anyway. With the addition of map information, a car will know the angle of a turn that's still 300 feet away. Navteq is in the process of collecting slope information, road width, and speed limits—all things that bathe the vehicle in more data than a human could ever handle.

Denaro believes that the key to making people comfortable with the shift from driver to planner will be the same thing that made pilots comfortable accepting autopilot in the cockpit: situational awareness. If a robot simply says it wants to go left instead of right, we feel uncomfortable. But if a map showed a traffic jam to the right and the machine listed reasons for rerouting, then we would have no problem pressing the Accept Route Change icon. We feel like we are still in control.

"Autopilot in the cockpit greatly extended the pilots' skills," Denaro says. Automation in driving will do the same thing.

Sebastian Thrun is standing in front of about 100 of his colleagues and teammates at a winery overlooking Silicon Valley. He has a glass of champagne in one hand and a microphone in the other, and everyone is in a festive mood. DARPA just gave Stanford a $2 million check for winning the desert race, and Thrun is going to use a portion of the money to endow the Stanley fellowship for graduate students in computer science.

"Some people refer to us as the Wright brothers," he says, holding up his champagne. "But I prefer to think of us as Charles Lindbergh, because he was better looking."

Everyone laughs and toasts to that. "A year ago, people said this couldn't be done," Thrun continues. "Now everything is possible." There is more applause, and then the AI experts, programmers, and engineers take small, conservative sips of the champagne. The drive home is curvy and dark. If only the party were happening in Thrun's future— then the champagne could flow unimpeded, and the cars would take everyone safely home.

Philip Smith

The Worst Date Ever for an Apple Tech

Life as an Apple Computer technician provides a
unique perspective on everyday life for this blogger.

Shortly after I returned to my hometown in 1997, I moved
to the local "big city" to try to make a living for myself.

I had always been into Apple Computers. Occasionally,
I repaired computers for a little extra money. When I first
moved to the big city . . . I started out selling Apple Computer parts on eBay . . . full time. (I still do so today.)
Through a few matters of circumstance, I became involved
with Apple Retail Representation and also was able to beta
test the PowerBook G3 Wallstreet. This was one of the first
laptops with a DVD drive. The main focus of my beta testing was for DVD video.

I loved it.

I had a CD player with a front stereo input installed into
my Geo Tracker . . . so I could listen to MP3's through my
car stereo. I was pretty far ahead of the curve . . . or at least I
felt like I was.

I was going to Atlanta (which was about 120 miles
away) three–four times a week. I was 23 and single.
HotORNot.com rates me at 8.5 . . . so I suppose you could
say I looked pretty suave.

Each trip to Atlanta, I was seeing a rather high profile client to do some Apple service work for them. I charged a very fair rate and sometimes took a colleague along with me. I included a dining and entertainment budget in my compensation proposals each time I made the 240 mile round-trip. Sometimes, I included overnight stays. Gas, at the time, in Atlanta was 79 cents a gallon.

For a while, I had been browsing through the personals at a site called MATCH.COM. Since I was traveling a lot to both Atlanta *and* Charlotte, I thought I would extend my range of searches to 150 miles. This meant that MATCH.COM would return results in the Charlotte, North Carolina; Knoxville, Tennessee; Greenville, South Carolina; and Atlanta, Georgia areas (and everywhere in between). Four states of girls to chose from.

I placed my own personal ad online. There were literally 1000s of results returned to me for my search criteria, which were:

Non smoker
5 feet 5 or taller
Athletic build
No children
Spiritual
Not a party animal
Not more than an occasional drinker

My personal ad was something along the lines of:

Let's cut to the chase . . . I have a lot of energy . . . if something such as ADHD actually exists . . . I have it. I like to be active. I like to run, and I love to play basketball. I like to hike, and I don't mind running upstairs instead of taking the elevator. Can you do any of that?

If you can't at least run across a mall parking lot to miss a rainstorm . . . stop reading.

As for an ideal date . . . my idea of an ideal date would involve the two of us making a nice picnic, carrying it to the mountains or to a local park, taking the convertible top off my Tracker, and putting the tarp over the entire car. We would take my laptop, put in a great DVD (yes it has a DVD), connect the audio to my stereo, and have our own private drive-in theater.

You don't have to attend church, but you gotta believe in God. A couple that prays together, stays together.

I like older women too . . . I have found in my brief dating experiences that women my age are unsure of themselves and a little too spunky. It's okay to have energy and be different . . . it's not okay to be completely different and obnoxious.

My longest relationship lasted five years . . . so I don't mind commitment, but I won't stay in a sour relationship either. When it's over, it's over.

I had a few responses and what I call "one date wonders." I'm not someone that sleeps around, so none of these dates involved sex.

One night . . . I get an instant message from a female in Atlanta. We hit it off pretty quickly. We instant message for right around a month. She tells me she is 28, 138 lbs, long blonde hair, and even sends pictures to back up the claims. (If you think you know where this is going . . . you're right . . . but bear with the story, because it gets interesting.)

After about a month of e-mails and instant messages, we decide to meet. I had been by a place called Dave and Buster's (D&B) several times while driving through Atlanta. So, I suggested meeting there . . . she said she loved D&B. It was close to another place I was visiting each trip called

MediaPlay, so I knew where it was. MediaPlay was one of the first places to carry DVD videos. I bought my first DVD there.

One key element to this story was just revealed to you. I am completely directionless. I have no idea where I am at any point of the day unless I have been there or driven there myself at least 20 times. The fact that I knew where Media-Play is in correlation to Dave and Buster's is key to understanding the nature of this story.

So, after my full day of work . . . I head to Dave and Buster's. I'm dressed in a nice pair of khakis, a white dress shirt, and my favorite tie. Yes, it's a King Kong tie!

We were to meet at 8 p.m. near the bar area of the main restaurant. I circled the bar noticing women who looked similar for about 45 minutes. No one stopping me, not noting anyone visibly 100 percent my girl.

At 8:45 p.m. I started to get the "I've been stood up queasy" feeling. I headed for the door. Right as I reached halfway through the parking lot . . . my cell phone rings.

"Rus . . . this is Paula . . . I am sooooooo sorry. I'm across town in the MediaPlay parking lot . . . my car is stalled . . . will you come get me?"

It didn't hit me at the moment, but remember how I told you I go to MediaPlay with every trip to Atlanta? That plays a key role in this story later.

Kind of feeling wishy washy . . . I agree and drive over to MediaPlay . . . which is about five miles away.

Let me back up for a moment . . . I was new to this whole Internet dating thing. I was kind of excited, and I really thought that I . . . *Mr. Tech Savvy* could make it work. I had this idea that of all the ways to meet women . . . this would be the way I would meet my wife.

I thought my car was cool but modest. I thought my attire was businesslike but casual enough. And . . . I was armed with my prototype Apple laptop with DVD player.

I also have to back up and say this next part is very atypical of me and my personality. People close to me know that.

So . . . I arrive at the MediaPlay parking lot . . . it was late October, but because of unseasonably nice weather, I had my convertible top down. I'm kind of weaving through the parking lot when a woman walks right out in front of my car. I almost hit her and have to slam on the brakes. I lean out of the window and say,

"Get out of the way bitch!"

You have to reference my frame of mind to actually know why I was so road raged. I had a long day of work, I wasn't in the most comfortable clothing, and I was slightly ticked off about being stood up.

The woman . . . who I almost hit says, *"Rus? Hey!"*

I say, *"Paula? Heyyyyy!"*

Not even missing a beat and I suppose not hearing or caring that I called her a bitch, she hops in the car and says, *"Well . . . give me a hug."*

Now, why did I say that . . . remember the description I gave you earlier?

Twenty-eight, 138 lbs, long blonde hair, and even sends pictures to back up the claims.

Well, this person is visibly 48 (or older), visibly 175 pounds or more, BLEACHED blonde hair (like with lemon juice and Clorox), those large Jewish woman sunglasses, and the worst of all . . . a man's plaid shirt . . . cinched at the waist with a cloth belt . . . and I suppose like a leotard type material over her legs and very high heels. She had bright red lipstick and honestly . . . looked nothing like her picture. Frankly, it was embarrassing to even be seen with her.

I almost had decided to just call it quits right then and there . . . but I was hungry and had not eaten in several hours. For me, who is a diet controlled ADHD . . . this wasn't good. Besides . . . I really wanted to check out Dave

and Buster's . . . and I just figured . . . I'm game enough . . .
it can't get any worse . . . I just won't see her again.

So . . . we drive to Dave and Buster's . . . we go to the
main restaurant . . . sit down and eat dinner. While browsing the menu . . . she elects to come nudge in on the same
side as me. I told her that made me feel uncomfortable. She
moves back to the opposite side.

While waiting for our food she reveals the following
details:

1) She is actually 46 . . . not 28.
2) She wanted me to call her by her online role-play
   gaming name: Athena.
3) She has five children ages: 2, 3, 4, 5, 7.
4) This is something she loves to do . . . meet people
   from online.
5) The pictures she sent were actually of her former
   roommate.

The details would not have bothered me so much
had she not lied and just told me up-front. I probably
wouldn't have gone out with her again, but I would have
at least gone out with her once . . . just for the simple fact
that we got along so well on the phone, in e-mail, and
instant message.

Plus, body type doesn't bother me, age doesn't bother
me, children don't bother me (as long as you can adequately
support them), and little quirks like . . . calling someone
Athena . . . well . . . I can't look past that . . . that's just kooky!

We finish dinner after I confront her about several of
these details . . . expressing that I was little upset that she
wasn't truthful about everything (or anything). She gives little mini whiny apologies.

I stretch my arms over my head and say . . .

"YAWNNNNNNN . . . it's late, I probably should head back home."

She quickly interjects with, *"But Rus, I thought you were going to show me how you win at Daytona USA?"*

Not one to ever decline to show someone cool technology . . . I agree to play a few video games.

Most modern video game arcades have moved away from tokens and quarters to a type of credit card that you swipe at each game to play. Dave and Buster's had a center kiosk to purchase these cards.

I walk up to the game card kiosk, and Paula (Athena) grabs me by the arm (lovingly) and says, *"Oh, I come here all the time . . . I already have a card."*

There were different levels of cards ranging from $5 to $20 and three premium level cards, $100, $250, and $500. Guess which one she had!

I know a little trick for the arcade version of Daytona USA . . . the kind where you sit in an actual cockpit with a steering wheel and the seat moves and rumbles on turns . . . and you compete against others in the same setup. Anyway, the trick allows you to turn into a racehorse. The race is instantly over for all drivers, and the racehorse car is shown on all the other players' screens as the winner.

So . . . I get in . . . throw my tie over my shoulder, and Paula (Athena) swipes her game token credit card in the slot. I select my car and start the race. I see Paula sort of walk away from my car and start to watch all the racers racing against me. I pull off my little trick, win the race, unfasten my seat belt, and get out of the car. Right as I am clear of the car, Paula races toward me . . . dives with her legs toward my waist and shouts,

"You won! You won . . . MY baby won!"

Now . . . for my build . . . I'm a pretty strong guy . . . but when a 175 pound hippo comes charging toward you and

you are in the path of the rampage there's little you can do
. . . other than fall backward. And . . . that's just what happened. In the middle of the floor . . . I fell flat on my back
. . . losing my breath.

I'm on the floor with my face turning red and gasping
for breath. With a raspy, choking voice I say, *"Paula get off,
I can't breathe."*

She's on top of me like a mechanical bull . . . wildly
cheering, *"I love you so much I just wanna eat you up"*
and:

"You won! You won . . . MY baby won!"

I'm looking up (flat on my back) . . . lack of air has made
my vision blurry . . . all I can see is this wild cowgirl bucking
on top of me and a crowd forming around the spectacle.
Momentarily, I catch my breath as someone in the arcade
pulls her off of me, indicating I was choking.

I get up . . . after being asked by a few people, if I was
OK . . . I walk away like nothing happened. The whole time
I could see out of the corner of my eye . . . faces of disgust and
stares of carnival sideshow freakishness . . . I could lip-read
people as I walked by them saying a very emotional, *"Oh my
God!"*

Again, why I didn't just abruptly end the date there and
just ditch her . . . well . . . I like to not think about that any
more.

So . . . we head over to the pinball games. I suggest we
play the pinball game Teed Off . . . just coincidence, not sar-
casm. At the time, I had this same pinball machine in my
apartment . . . I was quite good at it. When it came time for
a match, I was almost sure I would win a free game . . . I held
my hand up as it matched . . . fearing another episode.

After that . . . I stretch my arms over my head and say,
*"YAWNNNNNNN . . . it's late, I probably should head
back home."*

As we're leaving and heading back out to my car . . . she mentions that I need to take her back to her car at MediaPlay.

I must not have had my head on straight that night . . . Why didn't I realize something . . . why if she had her purse would she need to go back to a stalled car? Why would I just not take her home? She had mentioned that she lived only a few miles from our present location.

Well, obviously I didn't think of that detail. So, I carry her back to the MediaPlay parking lot. We're there in a few short minutes . . . on the way she's just going on and on about how well it's gone and how happy she was to have finally met me.

When I stop my car . . . she gets out and says . . . I'll be right back. I may be adding a little dramatic detail to her next action, but I could almost swear she went to the car just to think about things, possibly even make a quick phone call . . . she looked like she was sighing and breathing hard to build up courage for something.

This whole bit where I was waiting in the car and she was fidgeting in her minivan lasted a good 10 minutes. At this point . . . it was about 11 p.m.

I beep my horn.

She comes to the driver side and with a pouty voice and says, *"Rus, will you do me favor? See over there in the corner, will you take me there?"*

Her arm with straight pointed finger was pointing out the corner of the strip mall that contained MediaPlay. I had never noticed the little nook . . . with the little door that said, **RENT-A-SITTER.**

I said, *"What is that?"*

She said in a "you're a dumbass tone", *"What does it look like, silly? My kids are there, and we need to go pick them up."*

I say with a sigh, *"Okay"*

We drive to the corner of the parking lot; she gets out and goes inside. About 15 minutes later . . . she storms out of the front door cursing. In an insistent voice, she says,

"They won't let me have my kids!"

In a sort of double take way, I said, *"What! Whatta you mean they won't let you have them?"*

"They won't let me have my kids!" she repeats.

So, leaving the car running, I march inside. The kids are being prepared for departure with their coats and other accessories being draped on them from cubbyholes.

I go up to a large "I used to be a man" type woman and say, *"What's the deal? Why can't we have the kids?"*

RENT-A-SITTER Drill Sergeant says, *"Paula can't pay AGAIN!"*

I turn to Paula and huff and give her the evil squinty eye look.

Wanting to just end this nightmare . . . I ask how much it is.

RENT-A-SITTER Drill Sergeant says, *"$640!"*

Me, *"What!"*

I was outraged; I was actually prepared to shell out $80–$100.

After several attempts trying to run her credit card for almost any amount . . . I try to negotiate the price down. With not much budging I start to turn to the offensive . . . and inform the RENT-A-SITTER lady that they can't keep the children. She shows me a Georgia law about child neglect blah blah blah.

To make a long story even longer . . . basically I swiped my credit card for $580. That total was five children at $20 per hour for . . . get this . . . five hours and a $15 fee per child for picking them up after 11 p.m., and I believe I was charged something like $5 for something extra one of them ate. Somehow, this was discounted.

I slammed the signed receipt back on the counter and informed Paula I would be in the car.

Again, I suppose I should have just sped off. But . . . I waited.

Coming out of RENT-A-SITTER . . . the kids seemed like ants climbing an anthill as they piled one by one into my car. The littlest girl, I suppose the four-year-old, comes up to me and says in a cute voice, *"Thank you, Uncle Rusty!"*

In a disgusted but sarcastic kind of voice I say, *"I'm not your Uncle Rusty"* and help her climb with her siblings into the back of the car.

Paula follows.

At this point . . . details of reality start to flood my head. I say to Paula,

"We can't transport these kids without safety seats."

Her reply, *"Oh, I don't have enough for all of them."*

Now midnight and beyond . . . I suppose I'm just delirious at this point.

I say, *"Well, I'm not carrying them out in the open . . . it's close to Halloween; we'll put my tarp over them, and they can look like ghosts!*

The last thing I wanted to do was get a ticket for "lack of child restraint."

Oddly, she agrees that this is somehow a good plan. So, tarp over children . . . good . . . and we're off. Vent holes and tears in the tarp made eyeholes for some of the taller children. I tucked the tarp in tight around their bodies . . . it was a good costume effect.

We only had a few miles to go . . . but it was kind of nippy in the air at this point. The kids were whimpering,

"Mommy, I'm cold!"

"Mommy, it's dark!"

"Mommy, I'm scared!"

Perspective: these kids are in the backseat of the car in 55

degree weather with a 45 mile an hour wind, with a nasty tarp over their heads. If I was a child . . . I'd be scarred for life. I think I am.

We arrive at her apartment complex . . . I help unload the kids quickly.

I stretch my arms over my head and say with wide eyes and a sarcastic tone . . . *"YAWNNNNNNN . . . it's late, I probably should head back home."*

She interjects . . . like she was using **Kryptonite** on me and said,

"But Rus, I thought you were going to show me the DVD on your laptop? Show me just like you said you would in your personal ad . . . puhlease!!!"

Reference . . . at the time I had very few if any people interested in DVDs, and the multitude of portable players didn't exist as they do today. It was truly amazing . . . especially for a movie buff like me.

Still . . . I insisted no . . . "the date" was over.

With her whining in the parking lot and realizing that I might literally pass out from exhaustion (and possibly near death strangulation) if I were to drive home . . . I figured I'd play along and possibly get a small rest.

I tell her to be no more than 15 minutes putting the children to bed. I set up the car.

I plugged my Wallstreet G3 laptop into my car stereo . . . check. I set the Powerbook on the dash mount I made for it . . . check. I pulled the tarp from around the children and draped it over my Tracker . . . check.

Sort of dark in the car . . . the movie started. I only had a few movies . . . I decided on *The Game* with Michael Douglas and Sean Penn.

Titlewise . . . I wasn't thinking . . . I just liked the movie. Well . . . about 20 minutes pass. I'm in and out of consciousness (drift sleeping).

I want to stress a few things again. I don't sleep around. So if you think I was trying to salvage some sex out of this . . . think again. I also don't drink. So no alcohol influence was involved here either. I had just never thought anything like this was possible or could get worse. If there's one thing I have finally learned this year . . . it is this . . . it can always get worse, but usually the bad things in life are lessons in maturity.

Back to the story . . . after about 20 minutes . . . she comes back to the car, lifts up the tarp, closes the door, and quickly adjusts her seat reclined.

Right about that time, (on screen) Michael Douglas takes a blonde into his arms and starts to make wild love to her in a remote cabin.

Inspiration to Paula; she leaps over the center console and wedges herself between me and the steering wheel.

"What are you doing?" I say angrily.

"I love you so much I just wanna eat you up," she keeps repeating as she's wet kissing me with open mouthed dog tongued kisses.

I don't know what to do . . . I force my forearm to her throat. Which was hard to maneuver anyway. I keep saying, *"Paula get off of me."* I start to say it louder and with a little more pressure.

Which reminds me of another Michael Douglas film. Right then . . .

"KNOCK KNOCK KNOCK . . . open up, please!"
(on the window from outside through the tarp)

I suppose it sounded like wild lovemaking . . . the short wheelbase and weight of what was going on were rocking the car.

I figured it was the police. I started to whisper for Paula to get off of me.

I say, *"Who is it?"*

"KNOCK KNOCK KNOCK . . . open up, please!"
(on the window from outside through the tarp)
Again, *"Who is it?"*
"Just open up!"
(I hear from outside through the tarp and driver window)
"Hold on a sec!"
With the Titanic on top of me . . . I reach with my left hand for the window crank and roll down the window. I reach outside and pull up the tarp.

A guy says, *"Honey, when are you gonna bring him up?"*
I had had enough!!!!
With more strength and adrenaline than I think I have ever had in my whole life . . . I opened the door . . . unwedged her and literally threw her down into the parking lot. There was cursing in that somewhere I'm sure.

I threw my Tracker into reverse and sped out of the parking lot . . . running over the curb as I peeled out. My laptop flew off the dash; the tarp flew out into the parking lot like a parachute off my car.

I immediately called my best friend from high school to tell him the news. He couldn't believe it. I told him I was coming to see him . . . he agreed that I needed the therapy.

I drove all the way from Atlanta, Georgia, to Charlotte, North Carolina, at 1:30 a.m. that night.

I arrived at my best friend's house after having been on the phone with him for the four hour trip. I quickly fall asleep at his place and then head back home around 10 a.m. or so to my hometown about 100 miles away.

When I arrive back at my apartment, which was very close to an interstate . . . all over my door are Post-it notes:
"RUSTY, I LOVE YOU."
"RUSTY, I MISS YOU."

Spelled out on individual Post-its: *"R-U-S . . .*
I L-U-V U."
"RUSTY, MY LEG IS OKAY."
"RUS, MY PHONE NUMBER IS . . . ."
"RUSTY, I'M ACROSS THE STREET AT THE
MICROTEL HOTEL."

I decide to walk out of my place carefully (snooping and afraid) . . . I walk the long way to see the Microtel Hotel. Sure enough . . . same minivan from the night prior that was "stalled" . . . sitting right there in the parking lot.

I call the cops. I watch from across the street with a police officer as they question her in the parking lot. They radio over and ask me to positive ID her. I say, *"Yes that's her!"*

The police officer suggests I take out a restraining order. I do so (for $35).

Over the next two and a half months I receive daily e-mails from different free e-mail domains . . . same messages as on the door. Some of them clarifying details about "recruiting" people from the Internet for threesomes with her and her husband, the bump that I thought was "the curb" when I peeled out of the parking lot . . . was actually her leg, etc., etc.

I blocked as many of the e-mail addresses in my Yahoo account as I could. I eventually just switched to a slightly different spelling of my e-mail alias.

The End!
That was some good therapy. . . .
hope you enjoyed the story . . .
all true, with no exaggerations.

Justin McElroy

Defend Ironton

*How a decade-long gamble might just save a
small town*

Ironton, Ohio, may seem just like any other small town, but
those brave enough to look below the surface will find the
real, surprising truth:

Ironton really is just like any other small town.

It's got its own points of pride: the nation's oldest con-
tinually running Memorial Day parade and the Ironton
High Fighting Tigers, just to name a couple.

It's also got its problems. They all come back to one,
really: It's economically depressed, having lost nearly 1,500
jobs in the span of about 18 months. In a town of a little
under 12,000, that's not a downturn, that's a catastrophe.

The city council reacted the best way they know how,
trying to keep spending down and enacting a municipal fee
that's none too popular with the longtime residents. But
somewhere in a local basement, a group of gamers from this
area have formulated their own plan to save Ironton:
They're going to destroy it.

TickStorm is not like any other video game developer.
They're a developer with a clear mission: To surrender their
home to an alien onslaught in a game so popular that it will

single-handedly put Ironton on the map . . . and save their beloved city.

The year was 1999, and Baltimore native and Navy vet Rick Eid had just been relocated to Ironton by his employer, Cabletron. He'd been asked to start a training department for the networking equipment company—a new direction that quickly ran aground.

"I moved out here working for Cabletron's training department, and 10 months later, Cabletron shut down," Eid said. "But in that 10 months, I had really fallen in love with the area."

After years of moving around in the military, Eid had promised his two teenaged children, Rick Jr. and Nikki, that they could finish school in their new home. But Eid found keeping that promise to be difficult without work. Luckily, he was soon hired by Ohio University Southern (OUS), a branch campus in Ironton, which charged him with creating a game development department. It was an idea Eid bucked at, largely because he thought the coding would be too difficult for students but also because he wasn't very familiar with game design in the first place.

But never let it be said that Rick Eid is a quitter. He secluded himself in his office for a solid week, attempting to learn every in and out of a design program called 3D Gamestudio.

The classes filled quickly, but the new instructor discovered that his students were interested in something beyond an easy few hours of course credits. Eid found, as he taught, that they couldn't get enough. As their enthusiasm for projects continued outside the classroom, he hit upon the idea of creating his own game design company with the students, independent from the school. With few resources, no formal training, and practically no experience, the world's most unlikely game studio was born.

A STORM, SOME TICKS, AND AN IDENTITY

They happened upon the name almost by accident. They had already settled on Melee Games, before a quick Internet search showed it was already taken by several other companies.

Their next choice, the one that stuck, was a name from Eid's past derived from a female student who was trying to pick an e-mail identity during a particularly bad thunderstorm: TickStorm.

Oh, and also, the girl loved ticks. It's pretty much your typical company name origin story. But they figured it was memorable, and you wouldn't need a bit of Googling to figure out they were certainly the only ones using it. What the team still lacked was a big idea. They drew their inspiration, in the end, from the same economically depressed climate that had brought Eid into their lives in the first place.

"One of the reasons for picking game design to teach at OUS was that we wanted these guys who had high-tech skills to be able to do a job and not have to leave the area to be employed," Eid said. "And with game design, it's something you can do at home."

For the employees of TickStorm, home was Ironton, and it was a home they were willing to defend.

A MIRACLE GONE AWRY

The plot of "Defend Ironton!" begins like the answer to the city's prayers. A large manufacturing plant moves into town and employs all those that are struggling to find work. But the locals soon learn that the bosses of this new corporation (psst, they're actually aliens) have something far more sinister on their minds: abduction.

"They all start work, the doors close, and no one sees them again," Eid said. "They've put up this impenetrable field around the city, so the army can't come in; no one can. You're on your own, and it's up to the residents of Ironton to defend the city."

The agenda, besides the benefit of working with an area they're extremely familiar with, is to give Ironton the boost it so desperately needs; just a little bit of extra attention to help bring a real (hopefully nonextraterrestrial) economic savior to the town.

"The students love this area; they were born here; they want to stay here," Eid said. "Hopefully, we can put Ironton on the map."

TOTAL INSANITY

The team—now comprised of 44-year-old Eid and about eight of his students—didn't want to just slap the city's name on the box. They wanted authenticity, with all the town's buildings perfectly modeled, but reality soon intervened. After working for weeks to model the Depot, a now-defunct Ironton restaurant, the team realized that re-creating the entire city with as little experience as they had might have just been more than they could handle.

The group had limitless energy and passion but didn't have, as Eid said, a setting with no limitations, where they could "step out of reality a bit."

"One of the guys said, 'What if we put the game in an insane asylum? Think of the stuff you could do,'" Eid said. "We started brainstorming, spent an entire day doing nothing but storyboarding, and came up with so much fun for this game."

TickStorm's maiden voyage would be Insanity, an off-kilter, first-person shooter set in a mental institution. They don't have the money for top-notch rendering and lighting, so they're putting their faith in work ethics and their own creativity.

"The game play and humor in this are going to be a blast," Eid said. "Things like you come around a corner and a herd of squirrels starts attacking you; clowns walk by and wave and then walk into a wall. Every time you look into a mirror, you see a different reflection. It's total insanity!"

Although this may not be particularly rib tickling on the digitally printed page, Eid has enthusiasm to spare, and he manages to sell it. Besides, he's quick to add, Insanity (which they hope to release in 2007) is just a dry run for the big show, though it's a dry run that has to finance said show.

"We're learning quite a bit by doing Insanity," Eid said. "Whatever money we make from that, the group's already said they want to roll a good portion, if not all of it, back into the company so we can afford better computers for every one of them and better software. For instance, I have an Alienware laptop, too, that fried on me. I mean literally; smoke was rolling out of the keyboard."

COMING TO A TOWN NEAR YOU

With Insanity slated for next year, and Defend Ironton! due in 2010, times are tight, financially. But that doesn't deter TickStorm's big thinkers; in fact, Eid is already planning on a franchise.

"It opens the door to unlimited sequels; you know, *Defend Cleveland! Defend Cincinnati! Defend Baltimore!* The world's the limit," Eid said. "If we get to the point where we're big like Blizzard or like EA [Electronic Arts] with a graphics department, we can just continue to work on it."

For now, though, the going is slow. Most of the work is done on the weekends, not including that done by Eid, who recently left his teaching job to work on TickStorm full-time. The hope is that, one day, his whole team will do the same.

"The hard part about doing this on our own is that these guys have to have jobs; they have to work; some of them work at Pic 'n' Save and other places," Eid said. "They have to make money, so they can't spend all their time doing this. Not too many guys want to come and work for you when they're not going to get paid until the game sells. One day, we're hoping that these games sell enough that these are the only jobs they have to do and they don't have to work at McDonalds."

Eid himself has not yet drawn a paycheck from Tick-Storm.

MIGRATION

All of the long-range planning may seem far fetched, but Eid and crew don't see it that way. Their determination is almost fanatical. They're always working to improve their situation, whether it's the regular local area network (LAN) parties they put on for gamers in the area or small projects to help increase their tool set. For instance, they've even begun to pick up on Maya with personal learning editions, but they still don't have the money to buy it.

To that end, they've just picked up their first paying game design gig: creating a safety training game for the Southern Ohio Medical Center of Portsmouth. In the game, which the team is frantically building models for, players learn the proper way to evacuate the facility in case of a fire or other emergency. No, it's not Half-Life, but it's work.

The big games are still years away, but it almost makes the effort that much more noble. They're not just wagering their years of work on a game concept or play mechanics, they're wagering that, in 2010, there will still be an Ironton worth defending.

But TickStorm doesn't think that way, and neither does Ironton. In their minds, the game making a splash, and the city's rebirth is practically a forgone conclusion. This small southeastern Ohio city and the game studio share the same intangible power all the graphic artists and multimillion-dollar budgets in the world couldn't match: They believe.

Preston Lerner

This Is a Bike. Trust Us.

*And you won't believe how fast it goes or how
much the guy inside is suffering*

Barely visible against the vast asphalt expanse of the Nissan
test track, a white speck emerges from the soft light of the
Arizona dawn. As it approaches, it takes shape as what
might be a miniature submarine or maybe a giant supposi-
tory on wheels. Crammed within the tiny, fully enclosed,
artfully streamlined body is a world-class cyclist who's
reclining like a guy on a Barcalounger as he pedals furiously
enough to make his bike the world's fastest sweatbox. He
rockets past with a whoosh, and I suddenly understand why
his ride is called a human powered vehicle, or HPV, rather
than just a bicycle.

Whatever you call it, this little sucker is honking along
so fast that it could merge comfortably into traffic on the
405. Moreover, the rider plans to maintain this speed for the
next 52 minutes, thereby setting a world record by covering
nearly 55 miles in an hour without the aid of an internal
combustion engine, electric motor, or flux capacitor. Oh,
and we're not talking about a cycling legend like Lance
Armstrong, tearing it up on a gazillion-dollar bike built of
unobtainium. No, today's would-be hero is a 29-year-old

Brit by the name of Rob English who manages to race full-time even though he doesn't get paid for it. "I have a very cheap lifestyle," he explains, "and a very understanding mother."

Which brings us to the dirty little secret of cycling: The fastest, most innovative, highest-tech bikes in the world aren't found in the Olympics or the Tour de France. They're the creations of the small and largely mocked world of HPV racing, a close-knit community of freethinkers dominated by engineering geeks—many of them California dreamers—whose idea of bling is a platinum-plated pocket calculator. Even when ridden by top pros, conventional diamond-frame bikes rarely exceed 40 mph on level ground. Meanwhile, HPVs (also known as speedbikes) have blasted past 80 mph thanks to their sleek composite bodies and recumbent, that is, prone, seating position.

In October, HPV racers from all over the world will congregate near Battle Mountain, Nevada, for a weeklong series of late-afternoon runs along State Route 305. Each record attempt will entail four miles of banzai pedaling to accelerate up to top speed and then—when the cyclist is about to puke, pass out, explode, or all of the above—powering through a 200-meter-long timing zone. Four years ago, Canadian Sam Whittingham became the world's fastest human by blistering the speed trap at Battle Mountain at 81 mph.

On this Friday morning at the end of June, English and most of the other luminaries of HPV racing are braving the broiling desert heat here in Casa Grande to assault another world record, this one in the so-called Hour, the most hallowed mark in cycling lore.

The distance traveled in an hour from a standing start has been the ultimate test of a cyclist's skill and heart since

1876, when an Englishman riding a high-wheeler covered 15.8 miles over the grounds of Cambridge University. Five-time Tour winner Eddy Merckx called his record-setting Hour "the hardest ride I have ever done"—this from a cyclist so legendarily fierce that he was known as "The Cannibal." Whittingham, a former Canadian national team rider turned HPV superstar, set the record of 52.3 miles two years ago. Although he's not here to defend his crown, six of his rivals plan to take a crack at it on the banked 5.7-mile-long oval at the Nissan Technical Center North America.

The carrot they're chasing is the Dempsey/MacCready Hour Record Prize. In 1999, Santa Ana businessman Ed Dempsey and visionary engineer Paul MacCready of Pasadena offered $25,000 to the first cyclist to cover 90 kilometers, or 55.9 miles, in an hour. MacCready had achieved international celebrity with his Gossamer Condor, a human-powered airplane that won the Kremer Competition in 1977, and he thought another cash-money competition would inspire innovation in the hidebound bicycle community. As he watches English's run, MacCready acknowledges that he set the bar too high. So even if nobody breaks the 90-kph barrier this weekend, he's agreed to award the prize money after the final record attempt on Sunday to the three riders with the best Hour marks to date.

English's first lap, starting from a standstill, is 47 mph and change. His first flyer is better than 54 mph, and he backs that up with another lap at 52-plus. He's close to a record pace, and the crowd—about 50 HPV junkies—perks up. Then English's speed heads south. Fifty-one miles per hour. Fifty. Forty-eight. Because it's so early and there's some cloud cover, it's only 84 degrees. Outside, that is. Inside English's carbon-fiber cockpit, which is sealed with tape and could double as a thermos, the heat is brutal. His bike wiggles like a fish (pedal-induced oscillation, it's called) as he grinds away at the cranks. "It's an ordeal," says American record holder Matt Weaver. On Sunday, Weaver plans to ride his Cutting Edge II, an HPV so insanely complicated that it makes the Space Shuttle look like a Model T. You'd think he'd be pleased to see one of his competitors punking out. But the expression on his face suggests that he's imagining himself inside English's oven. "Once your body temperature gets up around 102 degrees, your muscles say, 'OK, let's wait until we cool down a bit,'" he says. "And there's nothing you can do about it."

English tries. Man, how he tries. But by hour's end, he's covered only 49.8 miles—a British record, though too slow to earn any Dempsey/MacCready booty. As he freewheels off the track, I hear him shout through the bodywork: "Get me out of here!" A furnacelike blast of heat eddies out of the bike as the canopy is ripped off. English is helped up and then crumbles to the ground, physically shattered. Covered with ice and draped with wet towels, he hardly moves for 15 minutes. "Dear God," he finally says, "it's unbelievable how hard that was. My heart rate was over 200 for at least 45 minutes."

Says John Weaver, Matt's father and a retired physician. "He went through hell. If he'd been a normal person, he'd be dead."

The architecture of the conventional diamond-frame bicycle hasn't undergone many fundamental changes since it was introduced more than a century ago, and for good reason. It's cheap, reliable, robust, easy to build, and remarkably efficient. But it could be better. Recumbent bicycles are faster, safer, and more comfortable. Build them out of space age composites and skin them in streamlined bodywork and they'll make conventional bikes look second rate.

A case in point: Brothers Steve and Craig Delaire race a

recumbent built by Steve, a motorcycle racer who went into the bicycle business in Santa Rosa because it was better for the environment. Middle aged and nobody's idea of world-class athletes, the Delaires aren't here to challenge the Hour record but to set personal bests. Even so, they go fast enough—Steve logs 43.3 miles and Craig covers more than 36—that they would have lapped Lance Armstrong if he'd been racing against them on a conventional bike.

The prospect of top pros being waxed by wankers prompted the Union Cycliste Internationale (UCI), the body that governs bike racing, to ban streamlining and recumbents back in the '30s. Cast out of the UCI family, 'bents have grown up as the unloved; overlooked; and, let's face it, peculiar-looking orphans of the bicycle world. According to biking stereotype, they're ridden either by old geezers with long beards and aero bellies or by over-the-hill engineers in bad shorts. Their cool quotient, in other words, is strictly negative.

Fittingly, the HPV movement was initiated by card-carrying members of the slide-rule club. Back in 1974, a streamliner designed by mechanical engineering professor Chet Kyle to energize his students at Cal State Long Beach set several speed records at the Los Alamitos Naval Air Station. In April 1975, he and Orange aerodynamicist Jack Lambie organized the first HPV event at Irwindale Raceway. MacCready served as official timer. "It was probably the strangest bunch of vehicles ever raced in one place at one time," Kyle recalls.

HPV racing has been relegated to footnote territory in recent years, tarred with the "clown cycle" brush. But back in the '70s and '80s, the mind-boggling speeds achieved by these unconventional racers briefly brought several of the studliest pros into the fold. U.S. Olympian "Fast Freddy" Markham was the first cyclist to exceed 50 and 60 mph.

Then, in 1986, he won the $18,000 DuPont Prize for breaking the 65-mph barrier. In recent years, Markham has limited himself to the top speed competition near Battle Mountain, but he has come to Casa Grande for what promises to be his last hurrah. "I'm not sure that I've got a record in me; I'm just an old guy." He smiles ruefully. "I sure wish I had this bike when I was in my prime."

At 49, Markham is taut and wiry, with a full head of graying hair but the restless energy of a teenager. He runs his hand over the sinuous body of his black beauty, lovingly polished by crew chief Gabe DeVault so that it resembles a patent-leather slipper. "Cycling is a matter of who's willing to suffer the worst," Markham says. "These days, if I get to hurting really bad, I'll just cave. I suffered through the pain years ago. Now my big thing is just finishing." Unfortunately, he can't even manage that. After turning a lap faster than 55 mph, his chain derails. Back in the paddock, he vows to try again tomorrow.

Next up is Damjam (pronounced Damian) Zabovnik, age 31, a soft-spoken Slovenian with a sparse beard who's working in an aircraft factory in northern California. Polish aviation student Jacek Kesy is his entire crew; they ferry their bike around on top of a Honda with a mismatched fender and 180,000 miles on the odometer. Last year, at Battle Mountain, Zabovnik set the European speed record of 73 mph—traveling backward. Yep, he sees the road in a mirror and steers via controls that are reversed. (Otherwise, he'd have to steer left to turn right.) Like Markham, he's fit and tightly sprung, but instead of having the Californian's sunny disposition, he's under a perpetual cloud of Eastern European pessimism.

Me: How's the track?
Him: Bad. Bad cracks in the turns.

Me: Can you break the record?
Him: The bumps are very bad. I adjusted the wheels
but . . . (*He mournfully shakes his head.*)
Me: So how's your bike now?
Him: It's better than nothing.

The originality, craftsmanship, and sheer nuttiness of Zabovnik's bike make it a favorite in Casa Grande. I'd come here expecting to find a vibe similar to land-speed car racing at the Bonneville Salt Flats, another venue where ingenious do-it-yourselfers chase records that interest nobody but them. But Bonneville mostly showcases blue-collar gearheads rooted in the glory days of hot rodding. HPV speedsters, on the other hand, are built by engineering iconoclasts looking toward a green future, and they're ridden by athletes who push the limits of human endurance. Zabovnik gets extra credit because he has a foot in each camp.

At the start, his knife-edged bike sways like a blade being brandished before hand-to-hand combat. He steadies himself and starts pounding out laps in the mid-50s. After 45 minutes, he's ahead of the record. Then we hear a tinny radio report from the chase vehicle: Zabovnik is down!

"I wasn't tired and I wasn't overheated," he says phlegmatically when he returns to the paddock. "But over the bumps, I was like this." He mimics a marionette being jerked around by a kid with a bad case of attention-deficit disorder. "The front tire exploded."

I expect him to be devastated: In 11 minutes, he would have been the world record holder. Instead, he looks resigned, as if this experience merely confirmed his conviction that the only immutable law of the universe is the one attributed to Murphy.

The last rider of the day is Weaver's crew chief Rob Hitchcock on one of Weaver's old bikes. A 43-year-old

Arcata general contractor who used to run the service department of a giant bike shop, he recently returned to school to study mechanical engineering. He is making his first run ever in what's known as a camera bike. To maximize aerodynamic efficiency, Weaver designs his HPVs without windscreens. Instead, riders "see" the outside world via video screens that display images from cameras mounted in the tail fin. Camera bikes are notoriously tricky to ride. Markham, the most experienced HPV cyclist on the planet, crashed his camera bike six times—and once ran over his former team owner—before reverting to a conventional model.

"You can do it, Rob!" Weaver shouts as he pushes Hitchcock away from the starting line. But Hitchcock is listing from the launch, and he swerves off the track and plops down in the infield. Weaver sprints over to right him, and Hitchcock tries again, riding not for a world record but simply to complete a lap around the track. His second start is better. But the video quality is so poor that he can't make out a bright orange pylon until it's a few yards in front of him, and when he swerves to avoid it, his bike crashes again.

There's a collective groan back in the paddock. All that time, all that effort, all that sacrifice, and for what? "It would be so easy to quit," Hitchcock says later, forcing a smile. "But I'm not giving up. This isn't the end of the world. I'm coming back. It was just so disappointing with all those people watching." He pauses to regroup. "In this sport, you've got to have a brilliant design, and you've got to have maximum performance. It's a combination of technology and athleticism. I'm an athlete of mind and body."

Hitchcock's father wraps him up in a bittersweet hug, and they pack up their gear.

Everybody convenes at the track at 4:45 a.m. Sunday to beat the heat, which is forecast to peak at 108 degrees. For the third consecutive night, the Cutting Edge crew thrashed past midnight, this time fixing the video system, the unique Weaver-designed front derailleur, and a way-cool water pump he built to bathe his head in ice water with each pedal stroke. "I think that's the most comfortable garage floor I ever slept on," Raymond Gage murmurs as he and the rest of the guys wearily start going over the bike yet again.

Weaver, 37, is the mad scientist who best sums up the HPV community's dual personality. A runner with Olympic aspirations until he blew out a knee, now an engineer who works on wind power, he's both a legitimate athlete and an honest-to-goodness geek. Tall, handsome, with perfect posture and a perennial smile, he's brilliant, charming, generous, approachable, polite, and articulate, a Gen X cross between Jack Armstrong and Tom Swift. "He's bloody clever," English says. "He knows everything about everything, and that's no exaggeration. I'm not stupid, and I can't keep up with him."

Weaver's flaw is that he can't leave well enough alone when perfection seems to be an all-nighter or two (or 10) away. The bikes that English and Markham ride, as well as the one Whittingham used to set the world record, are based on something Weaver built years ago but never raced because he'd already designed a new-and-improved model. "He's a genius," Markham says. "He revolutionized the sport. But he gets hung up on details. Sometimes you wish he'd forget about getting everything just right and say, 'We'll deal with that next year.'"

As befits his name, Fast Freddy is ready to go when the track opens for business. While Markham completes his first lap, English is wheeled to the starting line. Bike owner Dave Balfour, an Illinois optician, and his companion, Becky Aulenbach, a nurse, have retrofitted the HPV with several Rube Goldberg contraptions to keep the cockpit cool. But

English is still feeling the effects of Friday's heat, and the bike has some niggling problems. When his speed drops below 50 mph, he bails. "I wasn't going fast enough," he says, "so there was no point in killing myself."

Markham, for his part, has clocked the fastest lap of the weekend—55.6 mph—and he's cranking hard, carving a straight-edged furrow down the track. "I just love watching him ride," says his daughter, Tanya. "There's no wobbling at all. It's like he's on rails." Keeping time with a gigantic stopwatch that dates back to the dawn of HPV racing, Kyle is suffering through bipolar swings of joy (because Markham is smashing the world record) and worry (because Markham may not be pacing himself). Tanya, grinning, just shakes her head. "It's in the bag," she says with 10 minutes to go. Daughter knows best. Markham's official mark is 53.432 miles, more than a mile better than the old record. His face is dangerously gray when he rolls to a stop, and he looks as though he's about to burst into tears—of exhaustion rather than joy. "I don't feel that good," he croaks.

Incredibly, the Cutting Edge crew is still screwing around with Weaver's bike. The water pump won't stay fixed. First, it's secured with metal tape. Then with a hacksaw and baling wire. Then with a cordless drill. Then the drill bit disintegrates. Weaver remains imperturbably upbeat. "I'm encouraged in a weird way," he says. "Every time I've raced against Freddy over a long distance, I've beaten him. So if we're able to solve these little technical issues . . ."

While the bike is carted over to the starting line, still in pieces, Weaver warms up on a stationary bike as a friend fans him with a cardboard box. "Oh, the hour cometh," he announces in a mock-biblical voice. Over his head, he drapes a turquoise terrycloth shawl fashioned to absorb water from the pump. On top of it, he fixes a large piece of metal Mylar film—it looks like tinfoil—to retain the coolness next to his skull. The whole thing is secured with black pantyhose. It would be hard, frankly, to make yourself look more ridiculous.

Weaver climbs with a gasp into a trough filled with ice water to lower his body temperature. Next he strides across the paddock with water dripping off of him and the tinfoil sticking out from his head like Mercury's silver wings. At the starting line, the crew is still working on his bike. In a reluctant, regretful voice, race organizer Al Krause informs Weaver that he's got 20 minutes, tops, to start his run. Virtually everybody at he track has gathered around to watch Hitchcock and company feverishly try to put the bike together, thrashing like emergency room docs swarming over a flatlining patient. Weaver alone remains outwardly calm. "You're doing great," he says as he patiently waits to get started.

Finally the bike is ready. Weaver slithers under the carbon-fiber top tube and, after a series of painful contortions, worms himself into the cockpit. The fit is claustrophobic; encased inside the bodywork with a breathing mask affixed to his face, he looks as though he's being imprisoned within a fiendishly high-tech torture chamber. The body is screwed shut. As the seams are being taped, Weaver shouts in a weirdly disembodied voice: "A strap came loose! A strap came loose!" Hitchcock, with sweat rolling down his nose, doesn't even pause. "You've pretty much got to run with whatever you've got right now," he says.

The bike is rolled to the starting line. With a clank of the chain Weaver pedals off. After 50 yards, he slams down hard on his left flank. Amid general pandemonium, his crew wheels him back to the starting area for a second launch. For three days, this is the run we've all been waiting for, and the only thing that can stop Weaver now is Weaver himself.

He takes off again. He wobbles, veers left, right and back to the left, then stabilizes. At the starting line, there's a ragged, almost disbelieving cheer. On the back straight, the chase vehicle reports that Weaver is already up to 55 mph. In the paddock, there's a sense that this is a done deal. Two years ago, at California Speedway in Fontana, Weaver covered 51.4 miles in an hour despite a host of problems. If he's going this fast this quickly, he should be golden.

"That's Matt," says Cutting Edge crew member Carl Mueller. "The wire is always scratching him on the back as he goes under it."

Weaver's first lap is solid, but his second is slower than anticipated. And it soon becomes clear that he's nowhere near a record pace. Markham, lounging in his bike with his back to the track, starts to smile, and the tension in the paddock deflates. A handful of people are waiting at the finish line to greet Weaver at the end of his anticlimactic 48.7-mile run. But he doesn't stop, he doesn't even slow down, and the crowd scatters to avoid being bowled over. Weaver rolls to a halt a half mile up the track. When he's pulled from the cockpit, he's as wasted by the heat as English had been two days earlier.

Fifteen minutes later, he's back in the paddock. Despite disappointment, dehydration, and heat prostration, he stands in his bare feet and manfully explains what went wrong: The video was so bad that he couldn't see the people at the finish line until he nearly plowed into them. The mask slipped off his face before the start, so he had trouble breathing, and the water pump broke after a half mile, so he was being parboiled inside the cockpit. Oh, and nobody filled his water bottle, and since there wasn't time to install the GPS unit, he had no idea how fast he was going.

"Wearing that goofy headgear was like putting on two ski caps and climbing inside an insulated box," he says. "It was torture." Told that, at one point, he was doing 45 mph, he groans. "Forty-five miles per hour! Oh, that's pathetic!"

Fast Freddy is the story today. For 60 agonizing minutes, he turned back the clock and earned one last moment of glory, not to mention $18,000 in prize money. (Whittingham and Weaver are awarded $10,000 and $6,000, respectively, for previous Hours runs, while MacCready gives Zabovnik $4,000 and English $2,000 for being the two fastest foreign competitors.) "I was hurting," Markham admits. "I was really hurting. I think this Hour took a year out of my life. This ride was probably my greatest performance ever."

For Weaver, the event has been a study in frustration, but in defeat, he exhibits the character of a champion. "What we were trying to do required great precision, and great precision requires time and money, and time and money were two things we didn't have," he says. "Records are important, and it's disappointing not to break them. But with records, you're living in the past. At the core level, my interest is in the potential—what it will take to get us to the next level."

I catch up with Hitchcock at the hotel. "The first thing I'm doing when I get home is catch up on sleep," he says. And after that? He flashes a giant grin. "Battle Mountain is only 12 weeks away."

John Seabrook

Game Master

Will Wright changed the concept of video games with the Sims. Can he do it again with Spore?

In 1972, an engineer and former carnival barker named Nolan Bushnell started a video-game company in Santa Clara, California. As an engineering student at the University of Utah in the 1960s, Bushnell had become obsessed with an early computer game called Spacewar. The game's developers, a group of graduate students who were part of the Tech Model Railroad Club, at MIT, an early proving ground of computer hackers, had never considered selling the game; their idea was to demonstrate the appeal of interactivity and to take a first small step toward simulating intelligent life on a computer. Bushnell's ambition was more worldly. He wanted to manufacture coin-operated game-playing machines and license them to amusement arcades. He foresaw a new kind of midway hustle, in which the hustler would be inside the machine. "The things I had learned about getting you to spend a quarter on me in one of my midway games," he later said, "I put those sales pitches in my automated box." From this unlikely marriage—the computer lab and the carnival—the video-game industry was born.

The first product of Bushnell's company, Atari, was Pong, a simple, elegant game in which two players manipulated electronic paddles and sent a blip back and forth across a black-and-white screen. The game had two basic components. It was a simulation of table tennis, managing to render most of the game's rules, structure, and logic onto the screen. And it was an animation—a moving picture designed to complete the feedback loop between the eyes, the brain, and the fingers on the game controls. The game was designed by a former all-state football player named Al Alcorn, who was Atari's second employee. As Heather Chaplin and Aaron Ruby tell the story in *Smartbomb,* their recently published history of the industry, Bushnell took the handmade Pong game to Andy Capp's Tavern, in nearby Sunnyvale, and within weeks people were lining up outside the bar in the morning, before opening time, to play it. By 1974, Pong had made it to a pizza parlor in Hanover, New Hampshire, where I played it, and for the rest of that summer my dearest desire was to go back and play it again.

The games that followed Pong—Space Invaders, Asteroids, Missile Command, and Pac-Man, among others—were even more captivating, but the simulations remained the stuff of arcades and midways: sports, space aliens, zombies, shoot-'em-ups. In the 1980s, as the speed and storage capacity of computers and game-playing consoles grew, designers continued to improve the graphics. The simulation side of the games, however, never came close to realizing the Tech Model Railroad Club's old ambition of reproducing real-life dynamics on the screen. The best-selling video game this year is Madden NFL, in which you get to play pro football from the perspective of star players. Madden NFL is a far more sophisticated simulation than Pong was, but the content of the game is no closer to real life.

In the late 1980s, a new type of video game quietly

emerged—the God game. Computer animation is a brute-force project of converting graphic art into 2-D pixels and, more recently, into 3-D polygons, the building blocks of digital pictures. But to create a truly absorbing simulation, one that offers some insight into the nature of real life, is a much more difficult proposition. The designer must play God or at least the notion of God in Boethius's *Consolation of Philosophy*—a god that can anticipate the outcome of the player's actions and yet allows the player the feeling of free will.

Among the pioneers of the God game was Peter Molyneux, of Great Britain, who created Populous, in 1989. The game gives the player omniscient power over a variety of simulated societies. (You can help them or torture them as you wish, although your actions have consequences in the game.) Another important God-game designer, Sid Meier, has based his Civilization series, which began to appear in 1991, on historical processes, such as scientific discovery, war, and diplomacy. But the master of the genre—the god of God games—is Will Wright. Beginning in 1989, with SimCity, in which the object is to design and manage a modern city, and continuing with The Sims, in 2000, in which you care for a family in an ordinary suburban environment, Wright created situations that redefined the boundaries of what a game can be. "It occurred to me that most books and movies tend to be about realistic situations," he has said. "Why shouldn't games be?" To game designers, Wright is the Zola of the form: the man who moved the subject matter of games away from myth, fantasy, and violence and toward ordinary social life.

For the past six years, Wright has been working on a new game, which will be released in 2007. It is anticipated with something like the interest with which writers in Paris in the early 1920s awaited Joyce's *Ulysses*. At first, Wright called the project Sim Everything, but a few years ago he settled on the name Spore. The game draws on the theory of natural selection. It seeks to replicate algorithmically the conditions by which evolution works and render the process as a game. Conceptually, Spore is radical: at a time when most game makers are offering ever more dazzling graphics and scenarios and stories, Wright and his backer, Electronic Arts, are betting that players want to create the environments and stories themselves—that what players really like about games is exploring what Wright calls "possibility space." "Will has a reality-distortion field around him," his former business partner, Jeff Braun, told me. "He comes up with the craziest idea you've ever heard, and when he's finished explaining it to you the world looks crazy—he's the only sane person in it."

Wright's office is in a corner of a six-story building a few blocks from San Francisco Bay, in Emeryville, California. It has a balcony where he can smoke. The walls are covered with drawings in colored markers, which bear cryptic messages like "Star Map Issues." Wright, who is 46, is tall and skinny, with a long, narrow face and slender fingers. He dresses in more or less the same clothes every day—black New Balance sneakers, faded black jeans, a button-down shirt, a leather jacket, and thick aviator-style glasses. His skin is shiny and reddish-brown, in that way that a smoker's skin can look—half tanned and half cured. He sometimes has a wispy mustache and goatee. You don't really have a conversation with him; you mention an idea, and that triggers 5 or 10 associations in Wright's mind, which he delivers in quick bursts of data that are strung together with "um"s.

When I walked into his office, Wright jumped up and, after shaking my hand, said, "Here, try this, um, it's this really cool toy I found recently," and handed me a wireless controller for a small robotic tank that was sitting on the floor. It was facing another tank, which Wright was controlling. He started moving his tank around and shooting mine, watching me curiously, waiting to see how long it would take me to understand what was going on. I felt an odd tingling sensation in my hands, but I didn't pay any attention to it at first. Eventually, I realized that I was getting shocked: every time Wright's tank shot mine, an electric charge passed from the controller into my hands.

Wright had been working on a PowerPoint presentation of a talk he had been asked to give about Spore. "It's supposed to be about how I came up with the game, but what I really want to talk about is the history of astrobiology, so I'm doing both," he said. He moved over to the two computers in his office and clicked through some images, while describing the basic structure of Spore. At first, I was baffled. Up to this point in his career, Wright has been including more and more social realism in his games. But Spore is a surprise—at a glance, it looks like a "cartoony bug game," as one contributor to a gaming Web site put it. The buildings don't have the crisp urban lines of SimCity; they look more like the architecture in Dr. Seuss books. Wright has also introduced weapons and conquest. The violence isn't gratuitous—in some cases, you have to kill to survive—but it isn't sugarcoated, either. Not only do you kill other creatures in Spore, but you have to eat them.

At the first level of the game, you are a single-celled organism in a drop of water, which is represented on the screen as a 2-D environment, like a slide under a microscope. By successfully avoiding predators, which are represented as different-colored cells, you get to reproduce, and that earns you DNA points (a double helix appears over your character). DNA is the currency in the early levels of Spore, and as you evolve you can acquire better parts—larger flippers for faster swimming, say, or sharper claws for defeating predators. Eventually, you emerge from the water onto the second level—dry land—and your creature must compete with other creatures, and mate with those of your own kind which the computer generates, until you form a tribe. You can play a violent game of conquest over other tribes, or you can play a social game of conciliation. If you make clever choices, according to the logic of the simulation, you will survive and continue to evolve. Along the way, you get to acquire ever more powerful tools and weapons and to create dwellings, towns, cities. When your city has conquered the other cities in your world, you can build a spaceship and launch into space. By the final level, you have evolved into an intergalactic god who can travel throughout the universe conducting interplanetary diplomacy and warfare.

The images that Wright called up on the computer were supposed to illustrate the game, but they gave little sense of what it would look like. There was a slide that showed the equation for gravity, a slide about panspermia theory (the idea that life on Earth began with organic matter brought from space by comets and other "dirty snowballs"), and a picture of the cast of the early-1980s TV show *The Dukes of Hazzard*. Wright paused to say that, according to his calculations, based on the speed of radio waves, 150 stars have received *The Dukes of Hazzard* by now.

Spore isn't a multiplayer game, like the immensely popular World of Warcraft, which runs on "massively parallel" computers (a distributed system employing many networked machines); it's what Wright jokingly calls a massively parallel single-player game. If you enable an Internet feature, Spore servers will "pollinate" your copy of the game with content created by other players. In order to create the best content for your style of play—"the right kind of ecosystem for your creature," as Wright puts it—Spore builds a model of how you play the game and searches for other players' content that fits that model. If you create a hyperaggressive Darwinian monster, for example, the game might download equally cutthroat opponents to test you. In other words, while you are playing the game, the game is playing you.

Wright asked if I would like to try the Spore "creature editor," which is the first major design tool in the game. On the screen was a kidney-shaped blob that looked like Mr. Potato Head before you add the features. Wright showed me the menus for creating my creature's skeleton, body, eyes, and skin. I used the mouse to stretch the blob into a torso, changing the shape and length of the spine as I did so. I chose the parts from the left side of the screen—flippers, beaks, three-jointed legs—all of which would cost DNA points at this stage of the game. Wright explained, "You can choose different mouths—carnivore, herbivore, omnivore—which will determine not only how you will eat in the world but what type of voice the creature has." On the right side of the screen were graphics that showed the evolutionary advantages and consequences of each choice—speed, power, stealth, and so on. Switching to the paint menu, I applied a base coat of purple and then some orange stripes; the computer automatically shaded the colors, so that my creature's skin looked professionally textured.

"OK, now go to test mode," Wright said.

I clicked a button, and my creature sprang to life and started lumbering around the screen. It was a goofy-looking thing—a hammer-fisted apatosaurus with a potbelly, a long neck, and floppy dog ears. But it was a fully animated character, something that Pixar might have created, and I had made it in about three minutes. I felt as if I were playing with digital clay.

Electronic Arts is the largest producer of video games in the world, with more than 7,000 employees and with studios in

North America, Europe, and Asia. It makes or licenses software for many game-playing platforms, including computer games for PCs and Macs; console games for Nintendo, Sony, and Microsoft game boxes; handheld games for the Nintendo Game Boy (and the new Nintendo DS) and for the Sony PSP; and online games for playing on the Web. Most recently, EA has begun to make "mobile games," for playing on cell phones, a new and rapidly growing market.

EA was founded in 1982 by Trip Hawkins, a former marketing manager at Apple Computer, as "the new Hollywood," and it was at first supposed to be a haven for video-game auteurs. Hawkins proposed to treat designers, who had hitherto been regarded as mere engineers, as artists and to design sexy packaging that would evoke album covers, with the names of the creators emblazoned on the front. "Can a video game make you cry?" was one of the company's early challenges. Over the years, EA shifted its strategy toward games based on "proven content"—licensed stories and characters from film, sports, and TV, rendered in game form. (More recently, it has focused on creating its own intellectual property.) EA has also developed sports-simulation games, based on professional sports leagues, featuring the players themselves. As Steven L. Kent recounts in The Ultimate History of Video Games, it began in 1984, with Dr. J and Larry Bird Go One-on-One, a basketball game for which EA paid Erving and Bird to use their names and images. Since then, EA has created a sports-gaming empire. The latest version of Madden NFL, which was originally published in 1990, sold 2 million copies in its first week of release this August. In recent years, the company has acquired a Microsoft-like reputation for hard-nosed business practices—buying smaller development studios that can no longer afford the rising costs of game production and shutting out potential competitors with exclusive licensing deals.

The EA campus is in Redwood Shores, California, at the northern edge of Silicon Valley. Employees dress in shorts; there's a gym; the games in the company store are less than half price; and several meeting rooms are designed to look like sports bars. But, according to two class-action suits for "unpaid overtime," one filed by EA game artists and another by programmers, working for EA hasn't always been as much fun as it appears to be. Although both suits have been settled and EA has revised its overtime policy, during crunch times 80-hour weeks continue to be the norm.

While I was at EA, I was given a demonstration of The Godfather, one of the company's new games. You begin as a low-level criminal and attempt to become, through the clever use of violence and extortion, the head of the crime family. One of the game's innovations is that, in addition to killing opponents, you can also wound them by shooting them in the kneecaps or shoulders—and if you only wound them you can still extort money from them and thereby advance in the game. I also saw the latest installment of the Tiger Woods golf franchise. The golfer allowed EA to attach motion sensors to his body and face, and the data were rendered in computer graphics. The result is, among other things, a remarkable computer-animated version of Woods's famous smile—the way the upper lip slides up over the teeth is perfect. After hitting a good drive, you get to hear Woods whisper, "On the screws, Tiger."

After the demos, I met Larry Probst, the 56-year-old chief executive of EA, who started in the company's sales department in 1984. Probst explained that EA allowed Wright to put together a development team by choosing some of the most talented artists and programmers from EA's vast network of game makers. The company also constructed a separate headquarters for the 75-member team in

Emeryville, about 50 miles north of the corporate campus, near Orinda, where Wright was living. It was counting on Spore to help shore up its bottom line. The company's stock price had dropped almost 30 percent since April, and its sales figures were 20 percent lower than last year's. Probst blamed the company's problems on one of the cyclical downturns that hits the game industry every four or five years, when a new generation of gaming machines become available; this fall, both the PlayStation 3, from Sony, and Nintendo's Wii system go on sale. Traditionally, gamers stop buying the current generation of games in anticipation of those that will be developed for the new machines.

But there are reasons to believe that EA's problems are more systemic—indeed, that the entire game industry is on the verge of a fundamental restructuring. Not since the early 1980s, when video games began moving from amusement arcades into homes, has the future seemed so uncertain. While each generation of hardware offers the capacity for increasingly realistic graphics—like Tiger's smile—it also requires producers to devote more programming hours to filling that capacity. Twenty years ago, it was possible for one man to create an entire video game; today, development teams of 100 or more are the norm. Moreover, EA's basic product, which is a boxed game costing around $50, isn't as appealing as it once was. Many adult players prefer "casual games," which can be played on cell phones and in shorter sessions online. Instead of buying games at a store and bringing them home, customers want games they can get on the Web. Just as some in the film industry have begun to wonder about the economic feasibility of films that cost upward of $50 million to produce, so people in the game industry wonder whether big-budget games can survive in a climate that favors downloadable games that are cheap, short lived, and disposable.

During our conversation, Probst seemed most enthusiastic about the market for casual games, especially games for cell phones, which earned EA more than $100 million last year. "Think about what happens when 3 billion Chinese people have cell phones," he said at one point. But how do you convince a casual gamer, who is just looking for distraction, to play a game that is about evolution, city building, conquest, and interstellar travel? I asked Probst about this, and he said, "You tell people it's a Will Wright game."

Wright belongs to the last generation of game designers (and, indeed, human beings) who grew up before personal computers and game consoles existed. He built models of things as a kid: "ships, cars, planes—I loved to do that," he told me. When Will was 10, he built a balsa-wood replica of the flight deck on the *Enterprise,* which won an award at a Star Trek convention. He was also fond of the board games made by Avalon Hill, such as PanzerBlitz, a strategy game loosely based on tank warfare on the Eastern Front.

Wright's father, Will Sr., and grandfather were graduates of Georgia Tech's engineering school, and Wright keeps their graduation pictures hanging on a wall in his house, alongside a picture of himself. His forebears are crew-cut men in sober suits, about to embark on successful careers in making useful things. Then, there's Will Jr., who never graduated from college and who didn't fit into the family tradition—a gangly man-boy with a sweet, slightly stoned-looking grin. "Something went wrong with this one," Wright said, peering at the picture.

In the 1960s, Wright's father developed a new way of making plastic packing materials and started a successful company, which allowed the Wrights to live comfortably in Atlanta. Will's dad was also an excellent golfer. His mother, Beverlye Wright Edwards, was an amateur magician and actress. Wright flourished in the local Montessori school, with its emphasis on creativity, problem solving, and self-motivation. "Montessori taught me the joy of discovery," Wright told me. "It showed you can become interested in pretty complex theories, like Pythagorean theory, say, by playing with blocks. It's all about learning on your terms, rather than a teacher explaining stuff to you. SimCity comes right out of Montessori—if you give people this model for building cities, they will abstract from it principles of urban design."

In the evening, Will and his father would sit on the porch and talk about the stars, NASA's Apollo program, and the possibility of life on other planets. Wright was planning to be an astronaut, and his goal was to create colonies in space that would help relieve the pressure of overpopulation. His father thought this was a wonderful idea.

When Will was 9, his father died of leukemia, and his mother took him and his younger sister, Whitney, back to Baton Rouge, her hometown. Will went to Episcopal, a conventional prep school. He didn't like it as much as the Montessori school, although he enjoyed discussions about God with the faculty. "That's where I became an atheist," he said. He started at Louisiana State University when he was 16; two years later, he transferred to Louisiana Tech. He excelled only in subjects that he was interested in: architecture, economics, mechanical engineering, military history. He had impractical goals—in addition to starting colonies in space, he wanted to build robots. He dropped out again after two years; drove a bulldozer for a summer; and then, in the fall of 1980, went to the New School, in Manhattan, where he studied robotics. He lived in an apartment over Balducci's, in Greenwich Village, and spent a lot of time on Canal Street scrounging parts from the surplus electronics stores that used to line the street and using them to build a robotic arm.

In the spring of 1981, Wright answered an ad in a car magazine: Richard Doherty, a rally enthusiast, was looking for participants to compete in a point-to-point race between Farmingdale, Long Island, and Redondo Beach, California. Wright had a Mazda RX-7, which he and Doherty modified with a larger fuel tank and a roll cage. They wore night-vision goggles so that they could drive fast in the dark without headlights and avoid the cops. "Will said we should take the southern route, even though it was longer, because if we got stopped he'd be able to talk to the cops," Doherty told me. "We did get stopped in Georgia. We were doing 120, with no headlights, but it didn't take Will more than a couple of minutes to make the officer see why he had to let us go without a ticket." They won the race, establishing a new record of 34 hours and nine minutes.

After a year at the New School, Wright went back to Baton Rouge to live with his best friend. His family expected Will to take over the plastics company, but Will wasn't interested. (Eventually, they sold the business.) Souping up cars for rally racing was his main passion that summer, until his roommate's sister, Joell Jones, came to Baton Rouge for a visit. Jones was 11 years older than Wright; their families had been friends, and he had known her when he was a teenager. Now she lived in Oakland, where she was a painter and a social activist. She was back in Baton Rouge to recuperate after severing a nerve in her wrist. To extend the range of motion in her hand, Wright built a device out of metal and rubber bands. "Will would talk to me passionately about the need to colonize space, and I would say that it was more important to feed people on Earth," Jones told me. "Somehow we fell in love." When Jones went back to

Oakland, Wright asked if he could come and live with her; she agreed, on the condition that he didn't interfere with her painting. They married in 1984.

In the early 1980s, coin-operated machines began to decline in popularity, and home-video games began to take hold. Atari, which had popularized home-gaming consoles, was superseded by Nintendo, a venerable Japanese playing-card company, with its Nintendo Entertainment System (NES). As hardware, the NES was an improvement over the Atari machines (Atari's joystick controller was replaced with the directional "+" pad, which the player operated with his or her thumbs), but it was software, in the form of a Nintendo game cartridge called Super Mario Bros., that made Nintendo the industry leader. Shigeru Miyamoto, who had designed Nintendo's Donkey Kong for arcades, redesigned the game, changing the carpenter in the game, whose name was Jumpman, to a plumber, whom he called Mario, and adding a brother named Luigi and a far greater array of aids (golden coins, magic mushrooms), obstacles (fire-spitting enemies), and underground passageways, many of them drawn from Miyamoto's boyhood memories of exploring caves in the mountains near his home in Sonobe.

By the time Super Mario appeared, the syntax for game play was firmly established; it remains the standard grammar today. The player progresses through the game by defeating antagonists, restoring his or her energy with "power-ups" he or she finds along the way, accumulating bonus points to rise to progressively harder levels, many of which feature a "boss" who must be defeated in order to earn a "save game" and not have to repeat the level. Although Super Mario, which debuted in the United States in 1985, had a goal (to rescue Princess Peach from a giant reptile named Bowser), it also encouraged exploration for its own sake; in this regard, it was less like a competitive game than a "software toy"—a concept that influenced Will Wright's notion of possibility space. "The breadth and the scope of the game really blew me away," Wright told me. "It was made out of these simple elements, and it worked according to simple rules, but it added up to this very complex design."

In the late 1990s, Sony's PlayStation console replaced the NES as the dominant home game-playing system, and Microsoft's Xbox, introduced in 2001, is now the second-best-selling machine. But neither Sony nor Microsoft has had Nintendo's influence on basic game design.

In 1991, yet another phase in the game business began when a young programmer named John Carmack, who was, together with John Romero, a partner in a Dallas-based company called id Software, figured out how to program 3-D graphics for a PC, enabling the designers to give more depth to interior spaces and to create more realistic movements. According to *Masters of Doom,* David Kushner's 2003 book, when Romero first saw Carmack's 3-D program, he exclaimed, "This is it. We're gone!" Romero designed the graphics and game play for an ultraviolent game, which called on his own love of 1950s horror comics published by Entertaining Comics (EC), combined with a heavy-metal sensibility. The result was Doom, the defining first-person shooter, in which you play a "space marine" and the object is to kill the zombies that come at you as you advance deeper into Hell. Everything about the game was designed to inflame a teenage boy's fantasies of power while causing grave distress to his parents. In 1999, the elders' worst fears about the antisocial effects of first-person shooters seemed to be realized when Dylan Klebold and Eric Harris, the teenagers who massacred 12 of their classmates and one teacher at Columbine High School, in Colorado, were revealed to be obsessive players of Doom. Congressional hearings on violence in video games followed. More recently, the San Andreas version of the Grand Theft Auto series, in which the object is to pimp and steal your way to the top (you can get power-ups from mugging prostitutes), caused Hillary Clinton to cosponsor the Family Entertainment Protection Act, which would ban the sale of violent games to minors. Clinton also accused the makers of violent and sexually explicit games of "stealing the innocence of our children and making the difficult job of being a parent even harder."

One day in his office, Wright showed me an e-mail he had received from Lara M. Brown, a professor of political science at California State University, Channel Islands, in response to an essay he had written for *Wired* about the educational value of video games. Brown, who uses technology in her own teaching, wrote, "Most of us are in agreement that this younger generation—raised on video games—has learned to be reactive, instead of active, and worse, they have lost their imaginative abilities and creativity because the games provide all of the images, sounds, and possible outcomes for them. Our students tend to not know how to initiate questions, formulate hypotheses, or lead off a debate because they like to wait to see what 'comes at them.' They also have difficulty imagining worlds (places and/or historical times) unless you (as a professor) can provide them with a picture and a sound to go along with the words. . . . In essence, they seem to have lost the ability to visualize with their minds."

Wright, though, believes that video games teach you how to learn; what needs to change is the way children are taught. "The problem with our education system is we've taken this kind of narrow, reductionist, Aristotelian approach to what learning is," he told me. "It's not designed for experimenting with complex systems and navigating your way through them in an intuitive way, which is what games teach. It's not really designed for failure, which is also something games teach. I mean, I think that failure is a better teacher than success. Trial and error, reverse-engineering stuff in your mind—all the ways that kids interact with games—that's the kind of thinking schools should be teaching. And I would argue that as the world becomes more complex, and as outcomes become less about success or failure, games are better at preparing you. The education system is going to realize this sooner or later. It's starting. Teachers are entering the system who grew up playing games. They're going to want to engage with the kids using games."

Shortly after moving in with Jones, Wright began making a helicopter simulator on his personal computer (a Commodore 64). Eventually, the simulator evolved into a shoot-'em-up in which the player flies the helicopter over various cities and islands, trying to bomb buildings and blow up bridges. Wright showed the game to Gary and Doug Carlston, the founders of Broderbund, one of the earliest PC-gaming software companies. In 1984, Broderbund brought it out as a PC game called Raid on Bungling Bay, and it appeared as a Nintendo cartridge the following year. It was only a moderate success for the PC, but it sold a million cartridges, mainly in Japan, and because of Nintendo's generous royalty agreement with Broderbund, Wright says, "I made enough money to live on for several years."

In designing Raid on Bungling Bay, Wright noticed that he "was more interested in creating the buildings on the islands than in blowing them up." He started thinking of a game in which the point would be to design buildings or, maybe, to build a city. A neighbor suggested that Wright take a look at a 1969 book called *Urban Dynamics,* by Jay Wright Forrester, an MIT professor, which argued that urban planning could be carried out more rationally by a computer simulation than by humans, because the computer wouldn't be blinded by intuitive biases. In a later book, *World Dynamics,* Forrester laid out his proposal for a simulation that could manage the entire planet.

Computer simulations had been around since the 1950s, when military planners, climatologists, and economic forecasters began programming models of particular scenarios and dynamics and using them to predict outcomes. One early and well-known biological simulation was the Game of Life, created by a mathematician named John Horton Conway, in 1970. The game, which simulated the growth and death of a living creature, was based on the principle of "cellular automata," in which the programmer assigns simple rules to discrete units, or cells. It can be played on a plain 2-D grid, in which black squares represent live cells and white squares represent dead ones. Each cell reacts to the state of the cells around it. The rules are (1) any live cell with fewer than two live neighbors dies of loneliness; (2) any live cell with more than three neighbors dies of overcrowding; (3) any live cell with two or three neighbors lives; (4) any dead cell with three neighbors returns to life. Conway's purpose was to show how a simple structure of cells could be organized algorithmically to simulate complex, lifelike systems in which unpredictable or "emergent" outcomes occur.

Wright figured out how to combine Forrester's and Conway's ideas to imitate the dynamics of a city. The player would be responsible for adjusting around 100 variables in a way that allows the city to thrive. You establish transportation networks, power grids, hospitals, and schools. Each decision affects many other variables: a rising crime rate leads to a declining population, which erodes the tax base, which requires the cutting of some essential services—less funding for the hospital, for example.

Wright built a prototype of the game and worked on it for Broderbund, but the company could not see the commercial potential for a game you couldn't win. Eventually, Broderbund gave back the game's rights to Wright, and he set out to find a backer.

One night, at a pizza party in Alameda, Wright met Jeff Braun, a young businessman who was looking to get into video games. As Braun explained, "Will showed me the game and he said, 'No one likes it, because you can't win.' But I thought it was great. I foresaw an audience of megalomaniacs who want to control the world." Together, they founded Maxis, and they brought out SimCity in 1989. (Broderbund eventually joined the venture as a distributor; by then Wright had added a feature that allowed players to destroy their cities with various disasters—a volcano, an earthquake, an alien attack, a meteorite shower.)

SimCity was slow to catch on, but 17 years later the game has earned the company $230 million. A sizable number of players who first became interested in urban design as a result of the game have gone on to become architects and designers, making SimCity arguably the single most influential work of urban-design theory ever created.

In 1986, Wright and Jones had a daughter, Cassidy, and Jones made Wright promise to share the parenting equally so that she could continue painting. "He really did stick to that," she told me. "He spent a lot of time with Cassidy." While he was at home with his daughter, Wright began to turn over the idea for a new game, a kind of interactive dollhouse that adults would like as much as children. "I went around my house looking at all my objects, asking myself, 'What's the least number of motives or needs that would justify all this crap in my house?' There should be some reason for everything in my house. What's the reason?"

One morning in 1991, as Wright awoke in his house in the Oakland Hills, he thought he smelled smoke and called 911. Over the next half hour, the smoke got worse. "I thought, Uh-oh, this isn't trending well." He and his wife decided it was time to evacuate (Cassidy was away at a friend's house). They grabbed some family photos, jumped into Jones's car, and drove away. When they returned, three days later, the Oakland Hills firestorm had destroyed everything. Nothing was left except for some lumps of melted metal, the remains of their other car. In the months that followed, as Wright went about replacing his belongings, he started thinking about all the things people needed. "I hate to shop," he said, "and I was forced to buy all these things, from toothpaste, utensils, and socks up to furniture."

Three works helped Wright understand how he could turn these life experiences into a game. One was the book *A Pattern Language,* by Christopher Alexander and his colleagues at the Center for Environmental Structure, in Berkeley. The book identifies 253 timeless ways of building, which are classified as patterns—"Stair Seats," "Children's Realm," and so on—and it shows how these patterns can create satisfying living spaces. The idea is that the value of architecture can be measured by the happiness of the people who live in it. The second was the psychologist Abraham Maslow's 1943 paper "A Theory of Human Motivation," in which Maslow described a pyramid-shaped hierarchy of human needs, with "Physiological" at the bottom; and above it "Safety," "Love," "Esteem"; and, at the top, "Self-actualization." The third inspiration was Charles Hampden-Turner's *Maps of the Mind,* which compares more than 50 theories about how the mind works. Putting these works together, Wright formulated a model with which to "score" the happiness of the people in his dollhouse by their status, popularity, and success and by the quality of the environment the player designs for them—the more comfortable the house, the happier the people. Wright told me, "I don't believe any one theory of human psychology is correct. The Sims just ended up being a mishmash of stuff that worked in the game."

From a technical perspective, Wright's singular achievement in The Sims was to design a new kind of "object-oriented" operating system that modeled the complexity of social dynamics. As Chris Hecker, one of the developers on the Spore team, explained to me, "In Will's games, the objects themselves are encoded to interact with the environment around them. So if you introduce an espresso machine you buy from the online Sims mall, the Sims will be able to make espresso without having to reprogram the game. All you have to do is drop the object into the environment, and it will make other stuff happen. The objects create 'verbs,' as we say."

The original Sims had eight motives or needs— hunger, hygiene, bladder, comfort, energy, social, fun, and room— all of which are affected by objects in the world around them. Life for a Sim is the pursuit of happiness, but happiness depends on social interaction and consumption, and consumption requires money. For example, the cheapest bed in The Sims 2, which costs 300 "simoleons," brings your Sim one point of comfort and two points of energy; a 3,000-simoleon bed carries seven points of comfort and six of energy. Wright has said that he intended the game as a parody of consumerism, because "if you sit there and build a big mansion that's all full of stuff, without cheating, you realize that all these objects end up sucking up all your time, when they had been promising to save you time."

Almost no dedicated Sims player, Wright included,

actually follows the rules of the game, which force you to spend many hours working in menial jobs in order to be able to afford nicer stuff. Most players use the "cheats" that are widely available on the Internet and have been built into the game by the programmers. Cheats are short pieces of code you can type into the game that let you get around the rules. Typing "motherlode" into The Sims 2, for example, endows your Sims with 50,000 simoleons. But using cheats doesn't really feel like cheating, because playing The Sims doesn't really feel like a game. It seems more like gardening or fixing up your house. One of the game's small triumphs is to make work seem like fun. As my 14-year-old niece exclaimed recently, when I asked her what she liked about playing The Sims, "You've got one Sim who you've got to get to school, and another who needs to get to his job, and their kid has been up all night and is in a bad mood, and the house is dirty—I mean, there's a ton of things to do!"

When Wright took his idea to the Maxis board of directors, Jeff Braun says, "The board looked at The Sims and said, 'What is this? He wants to do an interactive dollhouse? The guy is out of his mind.'" Dollhouses were for girls, and girls didn't play video games. Maxis gave little support or financing for the game. Electronic Arts, which bought Maxis in 1997, was more enthusiastic. (Wright received $17 million in EA stock for his share of the company.) Wright's games are so different from EA's other releases that it was hard to imagine the two being united in the same enterprise. But the success of SimCity had already established Sim as a strong brand, and EA, which by then, 15 years after its founding, was becoming a Procter & Gamble–style brand-management company, foresaw the possibility of building a Sim franchise. Released in 2000, The Sims was an immediate hit; it went on to become the best-selling PC game of all time. EA has since licensed it to many other playing platforms and issues regular Sims "expansion packs," featuring new content, like Livin' Large, House Party, and Hot Date. (Wright worked on The Sims Online, which was a major redesign, but he has had nothing to do with the expansion packs.) The Sims franchise has earned EA more than $1 billion so far. EA's only misstep was The Sims Online, the multiplayer version released in 2002, which failed to attract the masses of players drawn to other multiplayer games, such as World of Warcraft and Runescape.

The Sims brought a huge new population to gaming— girls. That did not come as a complete surprise to Wright, since women made up 40 percent of his Sims development team and his daughter Cassidy, then 14 years old, had helped him tinker with the prototypes. When he was a kid, Wright told me, "I never played with dolls, which is more of a social thing than playing with trains—it's about the people in the house. Cassidy helped me see that. She and her friends got into the purely creative side of the game, rather than the goal-oriented side, which really influenced me a lot." Cassidy was traumatized to discover that the Sims could burn down their house, and die in the fire, if they weren't careful around the stove. Wright left that feature in the game.

An unintended result of The Sims' success is that Wright transformed the tactile experience of playing with dolls, which has been a part of children's development for thousands of years, into a virtual experience. The enormous success of The Sims means that children today can grow up without having the hands-on model-making experiences that Wright enjoyed as a child and that inspired him to make games in the first place.

One evening, I went with Wright to the house he and Jones moved into after the Oakland Hills fire. He drove a black two-door BMW with a fancy radar detector. The car was a mess, inside and out; Wright never washes it, because he wants it to look like one of the banged-up starships in *Star Wars*. Parking it in the garage, he led me into the house through a short hallway that was full of oddly shaped pieces of machined steel. Wright explained that these were left over from the days when he competed in gladiatorial robot contests called BattleBots, in which engineers attempt to build the most destructive remote-control robot vehicles possible. These ferocious machines fight in large Plexiglas boxes, ramming into each other at high speeds, trying to disable their opponents by flipping them over; the tournaments are like geek cockfights. One of Wright's robots, which he designed with the help of Cassidy, was called Kitty Puff Puff. It fought its opponents (which had names like the Eviscerator and Death Machine) by sticking a piece of gauze to its opponent's armature and then driving in circles around it, until the opposing robot was so cocooned in gauze that it couldn't move. Eventually, the organizers banned cocooning.

The house, a split-level, was on a hilltop in Orinda, and it had a lovely view of Mount Diablo in the distance. Jones's paintings—colorful, biomorphic abstractions—were hung on the walls, and in the yard were her sculptures: architectural-looking objects made of found metal. But Wright's stuff took up most of the space. Just inside the front door was the control console of a Soyuz 23 spacecraft, from the 1970s, which Wright bought from a former State Department official. Upstairs was his collection of unusual insects. Cassidy was away at college, but her prints—whimsical collages that feature drawings of rabbits and electric sockets—were also on display, and I saw a comic book she made, *The Adventures of Not Asian Girl*. On a porch off the living room were large blocks of alabaster that Wright was in the process of sculpting with hand tools into smooth, Brancusi-like shapes, a hobby that Jones had suggested to her husband as a way of expressing his artistic side. The rock dust and overflowing ashtrays on the porch suggested that he had been devoting a considerable amount of time to grinding stone lately.

The house was also filled with books. Some are what Wright calls "landmarks"—foundations for the design of one or another of his games. "Most of the games I've done were inspired by books," he told me. SimEarth, a simulation of the earth's ecology, was based on the Gaia hypothesis of James Lovelock, and SimAnt, an ant-colony simulation, was based on E. O. Wilson's *The Ants*. The key landmarks for Spore, however, were not books. They were Drake's equation and *The Powers of Ten*. The former, which he'd shown me on the computer screen in his office, is a formula devised in 1961 by Frank Drake, a radio astronomer, to estimate the number of possible worlds in our galaxy that might be populated with beings that could communicate with us. (About 10,000, according to Drake's calculations.) The latter is a short film by Charles and Ray Eames, made in 1977, which begins with a man lying on the grass in a Chicago park and then shows a series of images of the same shot, each taken from a position 10 times farther away than the last one, until the viewer reaches the limits of the universe at 1,024 meters (10 to the 24th power). Then it returns to the opening image and goes the other way, zooming into the man's skin, until at 10–16 you reach the limits of the inner world—the space inside a proton.

"I love *The Powers of Ten*," Wright said, "and I've always been a big fan of the Eameses. At the same time, I am really interested in the terms of Drake's equation, and when I began working on Spore I was using it to map some of the game play. At some point I realized that the terms of Drake's equation mapped neatly to the scale of *The Powers of Ten.* So I rolled the two up into Spore."

Wright seems to be more interested in making games than he is in integrating his ideas into a coherent philosophy. After you have played The Sims long enough, for example, you begin to recognize all the ways in which the simulation is not like real life. (The Sims 2, which came out in 2004, added more refinements to the basic design; in addition to the motives and needs, there are four different aspirations.) The Sims is only as realistic as the social theories it's based on, and these theories have been combined not according to scientific principles but for the purposes of entertainment. The Sims doesn't really model human dynamics; it merely gives you a model for exploring your own idea about how families work (just as playing with dolls does). Wright is not a visionary, in the sense that he is not the author of a world-view; he tailors his ideas according to the technical parameters of the simulation and the logic of games. Whether the game involves fighting intergalactic battles or modeling climate change, the simulation works according to a logic of its own. Wright may be the game industry's greatest auteur, but to a large extent he has abdicated authorship of his own creation.

Jones came home with some Mexican takeout, and we ate from the containers, in the living room. Jones is soft spoken, but she had a quiet authority around the house. She seemed a bit subdued. I asked her if she played her husband's games. "No, I don't. I'm not really interested in games," she said pleasantly. Later, she added, "Our daughter Cass used to say, 'We lived through the process of making the games, so we don't need to play them.' I think it frustrates Will that I don't play his games. Clearly, his games matter, on a deep level, to many people—take these online diaries people keep about their Sims. Wow. I don't know if they're avoiding their lives or learning about them. Me, I don't want to play a game to learn about myself." Several months later, when I heard that Wright and Jones were thinking about separating, and Wright had moved out, I recalled Jones's words.

I asked Wright if he was working on a new game. He said that, for the first time in his career, he was not. He was researching the Soviet space program and hoping to produce a documentary film about it. He said he was seriously considering a return to rally racing this November by competing in the Baja 1000—a race across the desert. (He later changed his mind.) He has a Hollywood agent and a TV development deal with ABC for a reality show exploring our relationship with technology in the home. But tonight, at least, he did not seem particularly engaged by any of these plans. With the prospect of Spore's launch ahead of him, he seemed a little lost.

In May, I joined some 20,000 members of the game industry—developers, marketers, distributors, buyers, press—in downtown Los Angeles for the industry's big trade show, Electronic Entertainment Expo, or E3. Electronic Arts, which had hoped in vain that Wright would have the game ready for the convention, was instead offering conventioneers the opportunity to see Wright demonstrate the game, inside a special Spore Hut that was set up next to EA's enormous pavilion.

By Wednesday morning at ten o'clock, when the trade floor opened, Wright was installed inside the hut, which could accommodate about 30 people. The line to get into the Spore Hut quickly grew to two hours long, snaking through the trade floor. Wright's mission was to play all the way through the game, which he estimates would require 79 years, if one played every aspect of it (Wright is designing cheats), in 17 to 20 minutes, over and over, for two days.

On the trade floor, screens showed guns, cars, football players, and Lycra-clad virtual babes, featuring "better breast shadowing, better breast physics, and deeper breast customization," as one gaming blog put it. It felt as if we were all inside some gigantic video-game machine—the place Nolan Bushnell had imagined 35 years earlier.

In the EA pavilion, I joined a clump of gamers watching each other play Battlefield 2142, the latest sequel to EA's popular shooter. I took a turn but kept breaking out in a sweat and being greeted with the alarming sight of my face reflected on the screen—scrunched up, red, demonic looking. Staggering out of the EA pavilion and into the cyber midway, I tried some of the nonviolent games, including Guitar Hero, in which the designers have ingeniously turned the controller into a guitar that you play by pushing buttons; it's like karaoke air guitar. I also tried SingStar Rocks! a PlayStation game that measures your pitch, phrasing, and timing and scores you as you sing. (Unfortunately, I chose Nirvana's "Come as You Are," which is not an easy song, and then compounded my problems by trying to sing in Kurt Cobain's register. "Awful" was the game's judgment of my performance, and it bothered me all day.) Finally, I got a demo of Left Behind, which is a Christian-themed video game based on the popular series of books by Tim LaHaye and Jerry B. Jenkins. You play a Christian in the streets of postapocalyptic Manhattan, and the object is to convert as many nonbelievers as you can before the Judgment Day. You get power-ups for finding bits of scripture, and praying raises your spirit level—which is represented by a graphic "slider" at the top of the screen. However, you have to kill hostile nonbelievers and can acquire some suitable weaponry for the job.

Violence drains your spirit level, but if you click on the pray button you can bring it up again.

The influence of Will Wright was not immediately obvious on the trade floor. His sandbox aesthetic is more noticeable in online virtual communities like Second Life, created by Linden Lab and based in San Francisco, which uses a similar operating system to The Sims Online. Second Lifers can buy space in a Sim-City-like community and use it for commercial transactions—conducted in virtual currency that can be exchanged for real money out in the real world. Aspiring musicians can perform onstage while their music is streamed over an audio channel. Second Life seems like a logical outcome of Wright's simulation games—and it isn't technically a game at all. When I asked Wright about Second Life, he said, "I think what you're going to see now on Second Life is people who will start to develop games—someone will invite other people to kick a soccer ball around, and it will go from there."

Wright's situation inside the hut was a little like his situation in the game industry—he seemed both enthroned and imprisoned. I half expected to find Bushnell at the door, charging people a quarter to see the geek. I could dimly make out Wright, seated behind a raft of computers on a raised platform behind the chairs, smears of color from the monitors reflecting on his glasses. There was a large screen on one wall, where the game was projected. He had been demonstrating Spore for about five hours straight when I got there, without lunch or cigarette breaks, although the EA handlers had brought him a Mocha Frappuccino, his favorite drink, from Starbucks.

I took my seat. There were little holes in the ceiling, with light behind them, to simulate stars. The walls of the hut were decorated with models of creatures that other players had designed. The lights went down, and the game began in the drop of water. "OK, so we start, and I'm trying to survive here—whoops, the guy wants to eat me." Wright narrated the game in the first person, and he seemed to be having fun. Using the creature editor, he put together a part-reptilian, part-avian creature, with yellow and purple stripes, four spindly legs, and talons at the end of its arms; it looked both cute and fierce. "OK, now I have to survive and eat—whoops, I'm going to run away from that guy. Whoops, not that way—this is a harsh world right now." His creature ate another creature's egg. "OK, now that I've eaten I feel like mating," and he located a mate the computer had generated for him. The creatures went at it, discreetly, behind a puff of smoke, to the sound of smooth jazz. "Procedurally generated mating," Wright said, with a smoker's chuckle.

Wright hurtled through the levels, evolution moving at hyperspeed as his creature acquired houses, tools, weapons, vehicles, and cities. While he was narrating his creature's adventures, Wright was also explaining how, in passing through the different levels of the game, the player would be progressing through the history of video games: from the arcade games, like Pac-Man, to Miyamoto's Super Mario, to the first-person shooters. At the tribal level you are playing a Peter Molyneux–style God game, and at the global level you are playing Sid Meier's Civilization. Finally, Wright reached the status of intergalactic god, with the power to visit other worlds. "Now we're going to go over to this place, which you can tell by the sliders has intelligent life on it, and this is actually a moon, a moon of this gas giant here. OK, this is an alien civilization, and there are a lot of different things I can do here diplomatically—I can actually use fireworks. OK, they seem to like that. Actually, now they're starting to worship me as a god. So I might decide to pick one of these guys up." A tractor beam came down from his spaceship and sucked up one of the creatures. The natives started shooting at him. "Oops. They were upset by that."

At a certain point in the performance, the crazy ambition of Spore became clear: Wright was proposing to simulate the limitless possibility of life itself. The simulation falls between Darwinism and intelligent design, into new conceptual territory. Wright had worked out the algorithm for life, as described by the philosopher Daniel C. Dennett, in *Darwin's Dangerous Idea*. Dennett writes, "Here, then, is Darwin's dangerous idea: the algorithmic level is the level that best accounts for the speed of the antelope, the wing of the eagle, the shape of the orchid, the diversity of species, and all the other occasions for wonder in the world of nature. . . . Can it really be the outcome of nothing but a cascade of algorithmic processes feeding on chance?" The old dream of the MIT hackers who came up with Spacewar—to re-create life on a computer—was coming true 40 years later, right here in the Spore Hut, in the form of a spindly, striped creature that looked a little like Will Wright himself.

After Wright's encounter with the other planet, he pulled back to reveal a vast galaxy of other worlds, some computer generated, some created by other players in the game who had reached the status of intergalactic gods— "more worlds than any player could visit in his lifetime," he said. As people in the audience gasped at the vastness of the possibility space, Wright's spaceship zoomed into the interstellar sandbox, looking for an uninhabited planet to colonize, just as young Will had promised his father he would.

Jeffrey R. Young

A Berkeley Engineer Searches for the "Truth" about the Twin Towers

Abolhassan Astaneh-Asl's computer simulations
suggest more-conventional skyscrapers might have
withstood the attacks.

When the World Trade Center's burning south tower crumbled to the ground five years ago, just 56 minutes after terrorists crashed a Boeing 767 passenger jet into its upper floors, Abolhassan Astaneh-Asl's horror was mixed with professional surprise.

As a professor of structural engineering at the University of California at Berkeley and an expert on steel structures, he thought that the buildings should have stood longer, even after such a catastrophic impact, and that the collapse should not have been so nearly vertical.

"From the day that I stood there and watched it collapse" on television, he says, "I was thinking that this is impossible. That there's something strange here."

Mr. Astaneh-Asl says he knew immediately that he wanted to be a part of the scientific response to the tragedy. He felt that his unique expertise could help in understanding how the two towers collapsed. He was well versed in the effects of terrorist bombings on buildings, having conducted research on blast effects after a car bomber brought down the Alfred P. Murrah Federal Building in Oklahoma City. And he had long studied how buildings responded to earthquakes and other natural disasters, so he knew that researchers must act fast to get the clues needed to understand what had happened.

The day after the attacks of September 11, 2001, Mr. Astaneh-Asl submitted an emergency grant proposal to the National Science Foundation asking for money to examine the steel at ground zero firsthand. Only days later, the request granted, Mr. Astaneh-Asl flew to New York and spent weeks at a recycling center where the towers' remains were being scrapped. There he inspected and collected samples of joints and other scraps of steel from what were once two of the tallest buildings in the world.

Though his NSF grant soon ended, questions about the collapse remained. Mr. Astaneh-Asl decided to create a computer simulation of the plane attacks, with as much detail as possible, in the hope that the unprecedented tragedy might yield lessons that could be used in the design of future skyscrapers.

He did not expect the discoveries he would make, the political obstacles he would face—or that, five years later, he would be involved in a struggle for skyscraper safety.

This week Mr. Astaneh-Asl is scheduled to present findings from his latest simulations in a lecture at Berkeley, making an argument that is sure to raise eyebrows in the engineering community and beyond.

If the World Trade Center towers had been built in a more conventional way and in strict accordance with New York City building codes—from which they were exempt because they were built under the auspices of the Port Authority of New York and New Jersey—the buildings probably would not have collapsed, he argues, and thousands of lives might have been saved.

Two extensive government investigations found no fault with the towers' structural design, and many engineers say that no engineer could have anticipated or shielded against kamikaze attacks by fuel-laden jetliners. And some say that what-if scenarios are fruitless lines of research.

Leslie E. Robertson, who helped design the twin towers, could not be reached for comment. But William Faschan, a partner at Leslie E. Robertson Associates, defended the buildings' performance. "It's extraordinary that any building could withstand that event and remain standing," he says. Mr. Faschan says he is not familiar with Mr. Astaneh-Asl's research and would not comment on its findings.

Even Mr. Astaneh-Asl is careful when discussing his findings, stressing that the people who perished in the buildings' collapse "were murdered by terrorists." But he insists that it is his obligation as an engineer to seek "the truth" about the buildings' history and structure. He speaks excitedly about the importance of his findings and talks eagerly for hours about the topic.

"We can avoid this happening to someone's loved ones in the future," he says.

"VIRTUAL REPLICA"

His work examining steel near ground zero gave Mr. Astaneh-Asl a few short minutes of fame. He was interviewed on *The NewsHour with Jim Lehrer,* CNN, and National Public Radio, as well as by several newspapers, including the *Chronicle.* At the time, he stressed the positive aspects of the twin towers' performance on the day of the attacks, noting that the length of time the buildings stood allowed most occupants to escape.

Mr. Astaneh-Asl says that from his inspection of the steel, he decided the collapse was not due to faulty welding or poor workmanship. That meant he was still not sure exactly how the collapse happened.

Then he got a call from an analyst from MSC Software Corporation, which makes high-end computer-modeling software used by carmakers and other businesses. Company officials had seen the Berkeley professor quoted in the news media and offered to donate the company's software for his efforts. He quickly took them up on their offer and began the next phase of his research.

Mr. Astaneh-Asl wanted the computer simulation to be as true to life as possible. That would require the blueprints and construction specifications for the twin towers.

"Basically you build the towers inside your computer with all the dimensions—you represent each element of your structure," he says. "It's really a virtual replica of your physical structure."

But the building plans for the towers turned out to be hard to come by, as the developers kept them sealed from public view. Experts say that most developers keep such documents private, viewing them as proprietary information, but Mr. Astaneh-Asl says he had hoped that, considering the circumstances, the plans would be made available to researchers.

He nearly got access by joining an investigation team led by the Federal Emergency Management Agency (FEMA) and the American Society of Civil Engineers, which brought together some two dozen researchers and engineers in late 2001.

Mr. Astaneh-Asl was initially asked to participate, but he says he was troubled that team members were all required to sign a nondisclosure form promising to keep certain details of the investigation, including the buildings' architectural plans, to themselves. Mr. Astaneh-Asl says he felt the agreement violated his academic freedom, and so he resigned from the team before its investigation got under way.

(The leader of that investigation, W. Gene Corley, says he believes the wording of the nondisclosure agreement would not have stopped any participant in the investigation from publishing academic papers about the structures. "It essentially said that we would not use information we obtained there to be used in a lawsuit against the owners and designers of the building," says Mr. Corley, who is senior vice president of the CTL Group in Skokie, Illinois)

After the Berkeley professor had nearly lost hope of obtaining the blueprints, he was invited to testify before the U.S. House of Representatives' Committee on Science, in March 2002, at a hearing titled "Learning from 9/11: Understanding the Collapse of the World Trade Center."

There, he was asked what "impediments" he had encountered in his research, and he replied that the largest one was his inability to get the design and engineering documents. Soon afterward, he was sent a copy of the plans by an official from FEMA. (He jokes that his wife wishes he had also asked for money to support his research.)

As Mr. Astaneh-Asl examined the construction documents, however, he was horrified by aspects of the design. He says the structure essentially threw out the rule book on skyscraper construction. "This building was so strange, and so many violations of practice and code were introduced," he says.

The design contains at least 10 unusual elements, he says. For example, rather than using a traditional skeletal framework of vertical and horizontal columns, the twin towers relied partly on a "bearing wall" system in which the floors and walls worked together to support each other, says Mr. Astaneh-Asl. That system allowed designers to use thinner steel in the buildings' columns and exterior than would be used in a traditional design, he says, adding that in some places the steel in columns was only one quarter of an inch thick. And he says the designers used stronger steel (measured in what is known as "yield strength") in some columns than is allowed by any U.S. building codes and that such steel is less flexible—and therefore more brittle—than the type traditionally used in such buildings.

As a result of such design elements, he argues, when the two airliners smashed into the upper floors of the towers, both planes plunged all the way in, wings and all. Airliners carry much of their fuel in their wings. His model clearly shows that in the initial fight between the plane and the buildings' exterior, the plane won, easily breaching the structure.

"It's like a soda can hit with a pencil," says Mr. Astaneh-Asl. "It was so easy that the plane went in without any damage and took the thousands of gallons of jet fuel in."

The structural innovations meant the developers saved money because they could use less steel, says Mr. Astaneh-Asl.

Efforts were made at the time of construction to verify the buildings could withstand anticipated forces, including high winds. The towers were among the first buildings ever to be modeled and tested in a wind tunnel before they were built. The buildings were widely praised for the efficiency of their construction.

But Mr. Astaneh-Asl argues that in engineering, innovations should—and usually do—emerge slowly, through evolutionary processes that follow time-tested practices. "Structural engineering is something that evolved," he says. "It was not invented."

"Unfortunately and tragically, when [this design] was subjected to this terrorist attack, there's no way this building could stand it."

Ross B. Corotis, a professor of civil engineering at the University of Colorado at Boulder, disagrees, arguing that the innovations did not mean that the towers were poorly designed. "I think our understanding of materials and our ability to analyze structural behavior gives us the ability to innovate more without introducing additional risk," he says.

Would a traditional structure have done better?

To try to answer that question, Mr. Astaneh-Asl and his team made another computer model in which they altered the design of the north tower's structure to make it more consistent with what the researcher calls standard engineering-design practice. Then he ran the same simulated plane into the structure in the same place it hit on September 11, 2001.

In that scenario, the airplane's wings are torn off, and therefore kept out of the building, when they hit the outer wall, while the fuselage still pierces the wall. "When it gets inside, there's not very much fuel," he says. Government reports found that it was not the damage from the planes but the subsequent fires that weakened the steel and caused the buildings' collapse.

Mr. Astaneh-Asl says he cannot be certain whether a more-traditional building would have survived the smaller fire that would have followed because he is not an expert on fires. Even so, he argues, if the World Trade Center towers had been designed "using the codes and traditional systems, the building most likely would have survived—it most likely would not have collapsed."

THE PREVAILING VIEW

Many engineers disagree with Mr. Astaneh-Asl's conclusions.

Mr. Corotis argues that "that particular design probably did better than most traditional designs would have done." One feature that helped, he says, was a cap across the tops of the towers that helped the buildings redistribute weight after the planes knocked gaps in them. Though he has not seen the blueprints for the structures, he says he is familiar with the buildings' innovative design, which has been widely publicized.

"Given the size of the planes and given the fire," he says, "the fact that they did come down is not surprising—but it's still shocking."

The most extensive investigation of the towers' collapse, completed by the National Institute of Standards and Technology (NIST), also found no fault with the structure, but it did acknowledge the buildings' unique history and design.

"The buildings were unlike any others previously built, both in their height and in their innovative structural features," a report on the investigation says. "Nevertheless, the actual design and approval process produced two buildings that generally were consistent with nearly all of the provisions of the New York City Building Code and other building codes of that time that were reviewed by NIST. . . . The departures from the building codes and standards identified by NIST did not have a significant effect on the outcome of September 11."

And the other government investigation into the collapse, led by FEMA, reached the same conclusion.

"We didn't really cite anything we thought the designers should have known about at that time," says Mr. Corley, who led that investigation. "It would have made no difference to what happened regardless of what building code it was built under."

Mr. Corley says Mr. Astaneh-Asl's simulations do not prove that the design was flawed. "If I know what's going to happen, I can design something that can do better" under those circumstances, he says.

Mr. Astaneh-Asl responds that his modified version of

the towers was not designed specifically for an airplane strike. "We designed this building assuming that they were building this building in 1970 following the [New York City] code without any consideration of an airplane," he says.

He sounds exasperated by what has come to be the accepted wisdom among engineers: that there was nothing wrong with the buildings. "I cannot see why the entire profession has agreed to sit in this convenient seat of saying that there is nothing wrong with our work," he says.

SKYSCRAPER SAFETY

For the most part, Mr. Astaneh-Asl has done little to publicize his findings so far, especially since he still hopes to publish a scientific paper about his latest simulation. He agreed to talk to the *Chronicle* only after a reporter called him to follow up on its previous coverage of his research.

He did present the findings in July at MSC Software's Virtual Product Development Conference, in Huntington Beach, California. An article that ran in *Design News* says the presentation had audience members "spellbound."

But Mr. Astaneh-Asl has been drawn into the political fight over the new Freedom Tower that is slated to be built at ground zero. Last year he joined an advisory panel of a group started by families of 9/11 victims. The group, the Skyscraper Safety Campaign, is lobbying to, among other things, require the new office tower to adhere to local building codes, rather than to the Port Authority's guidelines.

"They're going to design this building without going to City Hall and getting permits," says Mr. Astaneh-Asl, his voice rising. "Even if you want to change your kitchen, you have to get a permit."

The Berkeley researcher says he initially declined the

group's invitation to join because he wanted to remain completely independent. Aside from the free software, Mr. Astaneh-Asl says that his simulations research is not financially supported by anyone and that he and the graduate students who helped with the project have volunteered their time, even using their personal computers. He says he later agreed to join the Skyscraper Safety Campaign's advisory panel because he supports their argument about building codes but that the group has not given him any money.

Sally Regenhard is the leader of the Skyscraper Safety Campaign. Her son was a firefighter who perished responding to the attacks. Ms. Regenhard says that the government was slow to investigate the performance of the World Trade Center on September 11, 2001, and of the emergency response that followed. "There was a huge, huge force of don't ask, don't tell—don't ask questions," she says.

Many people outside engineering and government have developed their own theories about how and why the World Trade Center buildings fell. Some, wondering how buildings that easily withstood fierce wind gusts for decades were so quickly brought down by airplanes, argue that explosives planted before the attacks must also have been involved. Even some college professors have advanced such theories, though they have largely been dismissed (the *Chronicle,* June 23).

Mr. Astaneh-Asl also rejects such alternative theories. "I certainly don't buy into any of the conspiracy stuff," he says.

"Those are lightweight buildings," he adds. "There was no need for explosives to bring them down."

Kiera Butler

Listen to This

In little more than a decade Pitchfork *has gone from a labor of love to the Web's foremost musical tastemaker.*

Recently, the legendary *Village Voice* rock critic Robert Christgau told me one of his favorite facts about the music industry today: there are more hours of music recorded in a single year than there are hours in a year; it is literally impossible for one person to listen to everything. Even the most obsessive music zealots couldn't come close, but it's funny to picture them trying—a nation of rock-nerd zombies joylessly trolling the MP3 blogs, night after waking night. Nick Hornby meets *The Twilight Zone.*

For those who don't have 40 hours a week to devote to panning for gold in the vast muddy river of new releases, there's *Pitchfork* (www.pitchforkmedia.com), the 11-year-old Web magazine that does the sifting for you. The main thing that distinguishes *Pitchfork* from the *Rolling Stones* and *Spins* of the world is its focus: album reviews—five new ones every day—that are aimed at helping the overwhelmed listener. "In some other magazines, what this band or that band did on the road gets more words than 'Is the record good?' and 'Should you buy it?'" says Ryan Schreiber, *Pitch-*

fork's 30-year-old founder and editor in chief. Not so at *Pitchfork,* where the other features on the site—breaking news about the independent music scene, interviews with musicians, and features—are merely extras.

The *Pitchfork* staff *does* have 40 hours a week to spend filtering through new music, which is good because in their Chicago offices the first thing I noticed were the mail bins, stacked high against several walls and stuffed full of CDs. New releases come in at the rate of 500 to 700 every month from record labels, promoters, and musicians who hope to catch the staff's attention. It's not every day that the *Pitchfork* staff finds real gems in those bins, but when it does, the critical world listens.

Last year, *Pitchfork* was among the first to discover Clap Your Hands Say Yeah, an unknown, unsigned band from Brooklyn. After a *Pitchfork* reviewer in New York recommended it, Schreiber and the rest of the staff were floored by what they heard. In June, *Pitchfork* posted an enthusiastic review, praising the band for its "dizzily wowing vocal harmonies" and "richly buzzing" phrases. They rated the album a rare 9.0 on their 10-point scale, and in the months that followed, critics at major papers (the *New York Times,* the *Los Angeles Times,* and the *Chicago Tribune,* to name a few) weighed in. "*Pitchfork* is taken seriously," says the freelance rock critic Jason Gross. "To print critics, it's like, 'We're going to look like a bunch of old stupid sourpusses if we don't get in on this ASAP.'"

It's true. David Carr, the 49-year-old *New York Times* media columnist and pop-culture reporter, considers *Pitchfork* his best defense against becoming the dreaded Old Stupid Sourpuss. "I get records at work, and every once in a while I'll put them in, but I'm, like, a *dad,*" says Carr. "I can't be walking around my house with headphones on all the time."

As it has influenced everything else, the Internet is influencing rock music criticism, and in the rock criticism community, *Pitchfork* has become the first major Web-based tastemaker. Carr sometimes invokes *Pitchfork*'s opinions in his reviews. "In journalistic terms, *Pitchfork* allows you to express in short form that at least one tribe on the Web holds this band or that band in good regard," says Carr. "It turns everything into apples, if you know what I mean."

It's fair to describe *Pitchfork*'s founder, the baby-faced Schreiber, as the antisourpuss. He's infectiously friendly, and he can't sit still when he talks about music he loves. In fact, during the hours I spent at *Pitchfork,* he barely sat still at all. As a kid in Minneapolis, Schreiber was lonely. His parents, both real estate agents, were out a lot, and even in elementary school he immersed himself in music culture. "I was so into music I was kind of weird," he says. "Music was all I ever really wanted to talk about." As I talked with Schreiber and the conversation cycled again and again back to bands and albums, I realized that music is *still* all he really wants to talk about. I asked him what he liked to read. "All I ever read actually is music reference and music publications," he says.

In 1995, Schreiber was 19 and fresh out of a lackluster high school career, with no real desire to go to college. So from his childhood bedroom he launched *Pitchfork,* posting a few reviews every month and interviewing every band he could badger into talking. After a few months, he began updating the site daily, and for the next few years he worked part-time as a telemarketer in the evenings so he could build *Pitchfork* during the day.

Over the next 10 years, Schreiber expanded the site, soliciting freelance reviews and selling ads. After stints in a series of midwestern basements and bedrooms, *Pitchfork* made its above-ground debut in its new Chicago offices in the summer of 2005, when it became clear that the site—with a full-time staff of 5, more than 40 contributing writers, and 160,000 daily visitors—had finally outgrown the basement. Schreiber likes the new offices because "they remind me of an old detective agency." Next door to a law office in a small building in Chicago's Logan Square neighborhood, *Pitchfork*'s five-room suite is packed with thrift-store desks and squeaky chairs and a handful of no-frills laptops. No private offices; no conference rooms. Close quarters ensure lively discussion. (On the day I visited, someone said, "Let's hope the Red Hot Chili Peppers never put out another record"; someone else wondered aloud how such an "aesthetically unappealing band" could have gotten as big as it did, while others offered theories and explanations.)

Because *Pitchfork* has become a notorious arbiter of trends in rock, some readers have begun to wonder what the site's review-writing formula is—and whether there is a secret recipe an artist must follow to produce a *Pitchfork*-approved record. In 2004, Loren Jan Wilson, a University of Chicago undergraduate, went so far as to devote his thesis to the subject. Using a computer program he wrote, Wilson analyzed *Pitchfork*'s writing to detect patterns in reviews. He found that artists that sounded "sad" or "plaintive" often elicited good ratings, while those that reviewers described primarily as "confident" or "assured" were less likely to score high. Using what he learned, he wrote two songs designed to fill all the criteria for a favorable review, but that part of his research was inconclusive, as Schreiber and company never reviewed the songs.

On the day I visited the *Pitchfork* offices, I came ready to witness the reviewing formula in action. I watched as

Schreiber and Scott Plagenhoef, the site's 32-year-old managing editor, hunched over their laptops discussing the posting schedule for the next week.

"How big do we think the new Coup album is?" said Schreiber.

"It could B-List," said Plagenhoef. ("B-List" reviews appear second on the site's home page, under the featured review of the day.)

"Ughhh," groaned Amy Phillips, the 24-year-old news editor from the next room. "That record is terrible."

And the *Pitchfork* staff swears that this is how it goes. There is no formula except what seems too obvious: they pluck discs from the bins that they recognize from labels' release schedules and friends' recommendations, and sometimes they try a disc out just because its name or cover art appeals. The staff in the office discusses new releases, and *Pitchfork*'s writers weigh in from all over the country on a staff message board. After a bit of talking, writing, and listening, everyone has figured out how he or she feels about a record, and someone is in good shape to write a review.

As haphazard as the process seems, there's a lot of history behind the particular tastes of *Pitchfork*, a history that is inextricably tied to Schreiber's musical coming of age. In Schreiber's case, a lot of the changing a person goes through between the ages of 19 and 30—the widening of perspective and the solidifying of identity—happened through music. "When you're younger, there's a lot of resistance to listening to music the kids you don't like listen to," says Plagenhoef. "I think once you get a little older, that cliquish approach to listening to music changes." In the late 1990s, Schreiber began to pull away from the insular world of indie rock he once inhabited, where musicians produced records cheaply on independent labels. In that world, loudly despising the way corporate record labels commodified music was de rigueur. Sick of the strict indie code of ethics, Schreiber began to pay attention to the mainstream, and he was surprised at how much he liked Top 40 sensations like Destiny's Child and Gwen Stefani. From the echoes of electronic music he heard in pop songs to the Indian bhangra beats in the hip-hop that was popular at that time, he realized that mainstream music drew its influences from interesting and diverse sources.

But even as Schreiber's tastes were changing, *Pitchfork*'s readers remained deeply loyal to indie rock, and when *Pitchfork* began to cover the mainstream, readers resisted. Indie rockers are famously suspicious of anything that is commercially successful, and some readers assumed that *Pitchfork* was being paid by promoters to review popular music. Others, who couldn't quite believe that *Pitchfork* reviewers genuinely liked, say, the new Missy Elliot single, assumed the new coverage was, as Schreiber put it, "some sort of ironic flourish."

Schreiber knows that if readers suspected that he'd crossed the line from covering the commercial music industry to participating in it, he would lose credibility fast. Thus he is slow to discuss *Pitchfork*'s commercial success. In December 2004, because of a computer error, a Web page revealing the site's ad revenue became publicly available, and many readers were surprised at how much money the site was pulling in—and outraged that reviewers still earned only a measly $20 a review. When I asked Schreiber questions about finances, he squirmed and told me he didn't want to talk about the site's revenue because "that's not how we really define ourselves." But a current rate card reveals that the site is becoming increasingly financially competitive. Ad rates range from $3.50 to $8 per 1,000 impressions, with minimum buys of 100,000 to 300,000 impressions, depending on the size of the ad. In order for a medium-sized ad to appear 250,000 times on *Pitchfork,* for example, an advertiser would spend $1,250. (Web advertising rates vary widely—and they're often negotiable—but *Pitchfork*'s published rates are on the low side of average among sites that attract a young, hip demographic. They are similar, for example, to the rates of the media gossip site *Gawker.*)

Among older music critics, there's a kind of nostalgia for a time when albums didn't arrive in the mail every day, when finding out about new bands meant becoming friends with the record store clerks in the know. Before 1966, when a 17-year-old Swarthmore freshman named Paul Williams founded *Crawdaddy!* the world's first rock magazine, hardly anyone had even considered writing about rock; music journalists stuck to classical and jazz. One year after *Crawdaddy!* appeared, a young Jann Wenner started *Rolling Stone,* and in the years that followed, critics like Lester Bangs, Richard Meltzer, and Greil Marcus proved to the world that rock was worthy of analysis; they wrote popular music into the cultural consciousness.

Crawdaddy! folded in 1979, but *Rolling Stone,* like the Rolling Stones, is still at it. And according to Schreiber and Plagenhoef, the magazine, like the band, also looks old and tired. Around the *Pitchfork* office, a mention of *Rolling Stone* elicits a collective wince. "It's definitely an establishment magazine," Plagenhoef says. "It's the *opposite* of youth culture, which is what it's trying to cover."

In covering youth culture, the *Pitchfork* staff has the advantage of being overwhelmingly youthful. The senior citizens of *Pitchfork*'s stable of writers are in their mid-30s, and the youngest are teenagers. Besides the practical benefits of their age (late nights at rock shows are no problem; family responsibilities are minimal; bands often speak more candidly with a peer than with some old dude), Schreiber likes the passion that young writers pour into their reviews. If you browse the archive, that passion is hard to miss. Phrases like "on all levels, a total fucking triumph" and "fucking supersonic" are not uncommon ways to end a review.

But youthful passion has its limits. Although he enjoys *Pitchfork* critics' enthusiasm, Robert Christgau thinks their reviews tend toward "opinion-wielding for its own sake." *Pitchfork*'s writers, he says, simply aren't old enough to be able to put an album in its context, so they opine freely, blissfully ignorant of the past 60 years of rock history. "If these guys would like to leave their world, and especially go back in history, that's much harder. They just haven't heard enough music."

He has a point. Some of the reviews do seem to scream I-just-took-this-great-creative-writing-workshop-at-Bard, especially the ones that read more like prose poems than rock criticism. But free-form music reviews are nothing new. In his 1974 essay, "How to Be a Rock Critic," Lester Bangs wrote about the willingness of magazines to publish pretentious screeds. "Most of them will print the worst off-the-wall shit in the world if they think it'll make 'em avant garde!" he wrote. "You could send 'em the instruction booklet on how to repair your lawn mower, just write the name of a current popular album by a famous artist at the top of the cover . . . and they'll print it! They'll think you're a genius!"

While *Pitchfork*'s most opaque and pretentious reviews are a few notches of formal innovation short of a lawn-mower manual, Plagenhoef admits the writing can be "impenetrable and masturbatory." He's right. "Let's talk about reductionism, shall we?" begins the recent review of a Stereolab record. But the blather is far less common these days, especially because a few years ago, one particularly off-the-wall review contained so many factual errors that Schreiber had to retract it. You learn from your mistakes.

And from each other. Fundamental questions—like what makes a record a perfect 10—come up again and again in continuous staff discussions. Schreiber thinks a 10 is a "timeless classic," but Plagenhoef isn't sure "timeless" matters. "I don't think what someone might think about a record in 5 or 10 years should affect how we think about it today," he says. (On the perfect-zero front, the staff is more unified. They think about perfect zero the same way that Dante thinks about damnation: to deserve it, you have to do something deliberately foul. When The Flaming Lips produced an album that required four stereos and four copies of the album for a complete listening, the *Pitchfork* staff agreed to take the release as an act of hostility. It was a zero.)

Joe Levy, the executive editor of *Rolling Stone,* said of *Pitchfork,* "Those guys are working in the great, uncleared forest, free to grow without editing." That isn't exactly true. In the past several years, Schreiber and Plagenhoef have refined the editorial process, and writers sometimes go through several drafts before they post their reviews. But when unexpected problems arise, the staff makes decisions on a case-by-case basis, and sometimes those decisions draw criticism. Last year, Schreiber and Plagenhoef decided that the reissue of Neutral Milk Hotel's *In the Aeroplane over the Sea* album deserved more than the 8.7 rating they had given the original release. So they pulled the old review and replaced it with a new one—and a new rating of a perfect 10.0. Schreiber was surprised when readers complained. "I didn't think anyone would miss the old review," he wrote in an e-mail. "Turns out people will whine about anything."

Figuring things out as you go along isn't always comfortable or easy, but working in that great, uncleared forest also has its advantages. Schreiber has discovered that running *Pitchfork,* like finding great records, is a matter of listening widely and making up your mind, talking about your strong opinions with people who have strong opinions of their own, and then diving once again into your pile of unopened CDs and getting back to the real work of running the site. There's a lot going on at *Pitchfork* these days: the staff is planning the site's second annual music festival for July, and they're in the midst of expanding the site so it can support more MP3s so readers can do more listening. Throughout the day I spent at *Pitchfork*, Schreiber darted around the office, fielding phone calls and making plans. But every once in a while, he would look up at those towering crates of CDs and tell me guiltily, "I really need to go through those bins."

Top 10 Things I've Learned about Computers from the Movies and Any Episode of *24*

They couldn't show it if it wasn't true.

1. Megapixels aren't important: What determines the resolution of a photograph or audio recording is the "enhancement" algorithm run on it. Any image, when run through the proper enhancement, will reveal sufficient detail to recognize a face, read a license plate, and so on.

2. Computer screens output text at 4,800 baud and make chirping sounds while doing so: Sometimes, computers can be revved up to 9,600 baud, and sometimes, for instance when printing the names of conspirators, slow to 300 baud. There is a great deal of variety in the sound computers make when outputting text, though. It used to be a sound reminiscent of a line printer, but modern computers seem to implement a more "boop boop boop" approach. Oh, and most computers output in a 16 × 9 font.

3. All computer systems have back doors: Hackers can get into any system by way of "back doors" that are left by the people who originally designed the system. The password of the back door is generally the name of the programmer's daughter.

4. There are wire-frame schematics of every building on Earth: These schematics interface with a wide variety of sensor and alarm systems. They can be manipulated in real time and are infinitely zoomable (see #1 in this list).

5. Decryption works one character at a time, while the other characters cycle quickly through all possibilities: Face detection algorithms work the same way, as do most search algorithms. Oh, and every time a detail is revealed, the computer makes a beep. You know, really, most times a computer makes a partial computation, it makes a beep.

6. It takes 10 minutes to break into a system: A computer expert will first play coy, saying that they don't know how long it will take. Then, in a few minutes, they'll reveal that they can get in, but they'll need a few minutes more (damn it).

7. The U.S. government surveils the entire planet, in real time, and keeps the tapes: At first, this seemed implausible to me, but then I realized that they probably have a couple dozen Webcams in orbit and use enhancement.

8. People generally keep incriminating evidence in folders organized by code name: However, they often encrypt them (see #5). Oh, and computers erase data at 300 baud, in reverse (see #2).

9. Powerful people have Webcams that record from the middle of their displays: You and I dart our eyes back and forth between the Webcam and our own screen. Powerful people have intense video conversations while staring straight into the camera, and, therefore, the Webcams must be recording from the middle of their screen.

10. Powerful people have access to very powerful PDAs: The mobile computers used by powerful people not only support full video, they have access to the real-time wireframe schematics, decryption and enhancement algorithms, and so forth. Oh, and they can read any data in any device. Oh, and as phones? They have awesome coverage.

Farhad Manjoo

The Corporate Toll on the Internet

Telecom giant AT&T plans to charge online busi-
nesses to speed their services through its DSL lines.
Critics say the scheme violates every principle of
the Internet, favors deep-pocketed companies, and
is bound to limit what we see and hear online.

To say that AT&T was once the nation's largest phone
company is a bit like describing the Pentagon as America's
leading purchaser of guns and bullets. Until its government-
imposed dissolution in 1984, AT&T, which provided a dial
tone to the vast majority of Americans, enjoyed a market
dominance unlike that of any corporation in modern history, rivaling only state monopolies—think of the Soviet air-
line or the British East India Tea Company—in size and
scope. In commercials, the company encouraged us to reach
out and touch someone; the reality was that for much of the
20th century, you had no choice but to let AT&T touch your
loved ones for you.

Now—after a series of acquisitions and reacquisitions so
tangled it would take Herodotus to adequately chronicle
them—AT&T is back; it's big; and according to consumer
advocates and some of the nation's largest technology companies, AT&T wants to take over the Internet.

The critics—including Apple, Amazon, eBay, Google, Microsoft, and Yahoo—point out that AT&T, along with Verizon and Comcast, its main rivals in the telecom business, will dominate the U.S. market for residential high-speed Internet service for the foreseeable future. Currently, that market is worth $20 billion, and according to the Federal Communications Commission (FCC), the major "incumbent" phone and cable companies—such as AT&T—control 98 percent of the business. Telecom industry critics say that these giants gained their power through years of deregulation and lax government oversight. Now many fear that the phone and cable firms, with their enormous market power, will hold enormous sway over what Americans do online.

Specifically, AT&T has hinted that it plans to charge Web companies a kind of toll to send data at the highest speeds down DSL lines into its subscribers' homes. The plan would make AT&T a gatekeeper of media in your home. Under the proposal, the tens of millions of people who get their Internet service from AT&T might only be able to access heavy-bandwidth applications—such as audio, video, and Internet phone service—from the companies that have paid AT&T a fee. Meanwhile, firms that don't pay—perhaps Google, Yahoo, Skype, YouTube, Salon, or anyone else—would be forced to use a smaller and slower section of the AT&T network, what Internet pioneer Vint Cerf calls a "dirt road" on the Internet. AT&T's idea, its critics say, would shrink the vast playground of the Internet into something resembling the corporate strip mall of cable TV.

The fears have been deepened by AT&T's new heft. Early in March, AT&T announced that it will spend $67 billion to acquire BellSouth, the phone company that serves nine states in the Southeast. The merger will make AT&T the nation's largest telecom company and the seventh-largest corporation of any kind. According to one study, the new AT&T will take in almost a quarter of all money American households spend on communications services. In addition to maintaining a near monopoly on local phone and DSL service in 22 states, the new AT&T would provide land-line long-distance service throughout the country; cellular coverage through its subsidiary Cingular, the nation's largest wireless carrier; and soon, even television broadcasts to millions of Americans.

The government is expected to approve the AT&T-BellSouth deal, but the merger has already prompted debate in Congress and at the FCC over how this new behemoth may control content online. Currently, there are few rules governing what broadband companies can do on their network lines; if AT&T wanted to, for instance, it could give you only slowed-down access to the iTunes store unless Apple paid it a cut of every song you buy.

To fight back, online companies like Apple and Amazon, along with Internet policy experts and engineers, are pushing the government to draw up a set of rules to ensure what they call "network neutrality." The rules, debated this past February in a Senate hearing, would force broadband companies to treat all data on the Internet equally, preventing them from charging content companies for priority delivery into your house. AT&T and other broadband companies oppose laws to restrict how they operate online—the free market, they say, will ensure an even playing field. In 2005, phone companies poured nearly $30 million into lobbying to ensure that the telecom industry remains free of regulation.

The battle may sound wonky, but its outcome could well determine the shape of tomorrow's media universe. Increasingly, we're all using the Internet for much more than surfing the Web; film, music, TV, and phone companies are looking at the network as the primary channel for delivering media into our homes, and AT&T and other tele-com firms are spending billions to deploy deliciously fast fiber-optic lines to handle the expected traffic. The regulatory tangle between broadband providers and Web companies over network neutrality reflects a more fundamental fight over precious communications real estate—a battle for control of the lines that will serve as our main conduit for media in the future.

Each side predicts dire consequences if its opponents win. Jim Ciccone, AT&T's senior executive vice president for external affairs, says that if broadband service is regulated, AT&T won't be able to recoup its costs for building these new lines—"and then we don't build the network." The Web firms say that if the big broadband companies are allowed to charge content firms for access to your house, we'll see the Internet go the way of other deregulated media—just like TV and radio, where a small band of big companies used their wealth to swallow up consumer choice. If broadband companies get their way, says Jeff Chester of the Center for Digital Democracy, the Internet will one day feature nothing much more exciting than "the digital equivalent of endless episodes of *I Love Lucy.*"

In 2003, when Internet policy experts first began discussing network neutrality, their primary worry was that broadband providers would strike deals with certain Web sites to block people's access to competing sites or services online. For instance, what if Comcast worked with Barnes and Noble so that every time a Comcast Internet user pointed his or her browser to Amazon.com, he or she was instead redirected to BN.com? FCC officials have frowned upon the possibility of ISPs blocking certain Web sites, but they have not regulated against it; Paul Misener, the vice president for global public policy at Amazon.com, argues that "under current rules," a company like AT&T "would be able to block us without punishment."

Although such actions are theoretically possible, most experts concede that broadband firms wouldn't do something as brazen as blocking customers from going anywhere on the Web; such actions would probably prompt immediate regulation. Now Amazon, eBay, Google, Yahoo, and others argue that broadband firms like AT&T, Verizon, and Comcast are looking to institute a more subtle kind of discrimination. They're looking to "prioritize content from some content companies over others," Misener says.

In fact, AT&T is not at all secret about its plans. In an interview with *BusinessWeek* magazine last year, Edward Whitacre, AT&T's CEO, took a hard line against Web companies that oppose paying for high-speed access to AT&T's customers. "What they would like to do is use my pipes free, but I ain't going to let them do that because we have spent this capital and we have to have a return on it," he said of Google and Microsoft. "Why should they be allowed to use my pipes? The Internet can't be free in that sense, because we and the cable companies have made an investment and for a Google or Yahoo or Vonage or anybody to expect to use these pipes [for] free is nuts!"

The pipes Whitacre is referring to are those his company is building under a plan it calls Project Lightspeed, a multibillion-dollar initiative to install high-capacity fiber-optic Internet lines into thousands of residential neighborhoods across AT&T's service area. The company expects to serve about 18 million households with fiber-optic lines by 2008; Verizon has similar plans to roll out fiber lines. The new pipes will dramatically improve Internet speeds to home customers. Today a typical DSL line downloads data at about 1 or 2 Mbps, and cable modems run about double that rate. Advanced fiber-optic systems will see download speeds of at least 25 to 30 Mbps. Today's DSL can barely download a single standard-quality video stream in real time. In tests AT&T recently ran in San Antonio, Project Lightspeed lines carried three standard-quality streams and one high-definition stream down the line simultaneously.

What will customers do with all this broadband capacity? As the phone companies envision it, we'll use it to watch a lot of TV. Both Verizon and AT&T are betting heavily on a technology called Internet protocol television (IPTV), a service that delivers television signals into people's homes over the new fiber-optic Internet lines. According to the phone companies, IPTV will be a boon to consumers, delivering high-quality video and advanced services like TV shows "on demand" and providing much-needed competition to cable companies.

What's not clear, though, is what else—besides watching TV—customers will be allowed to do with the new lines. This is the heart of the fight over network neutrality. If you subscribe to AT&T's Project Lightspeed service, will you be able to use the 30 Mbps line coming into your house for, say, downloading high-definition movies from Apple, high-definition home videos from YouTube, or some other bandwidth-heavy application we haven't yet dreamed of? Or, instead, will AT&T reserve the line for its own TV service and for data from other companies that pay a fee— thereby making AT&T the arbiter of content in your home?

At the moment, phone companies are cagey about their plans. What they will say is they're not going to stop their customers from getting to any site or service on the Internet. "Let me be clear: AT&T will not block anyone's access to the public Internet, nor will we degrade anyone's quality of service," Whitacre said in a speech to a trade conference in Las Vegas recently. "Period. End of story." But just because AT&T won't block people from accessing Google's videos doesn't mean it will give Google's videos the same status on the broadband pipe as other content—meaning that while AT&T's TV service may come in at high-definition quality, those from competing firms might only run at standard definition.

Indeed, AT&T and other network operators are building their networks in a way that would make it possible to split up network traffic into various lanes—fast, slow, medium—and then to decide what kind of data, and whose data, goes where, based on who's paid what. Broadband companies argue that engineering their networks in this way will benefit customers in two ways. First, they say, splitting up the Internet into several lanes will generally improve its efficiency—the network will simply run better if it's more logically managed.

The phone companies' second argument concerns cost. If AT&T builds a blindingly fast new Internet line to your house but only allows some firms—firms that pay—to get the fastest service, it can significantly offset the costs of the build-out. And that's good for you, AT&T says, because if the company can charge the likes of Apple and Google to pay for the line, it doesn't have to charge you. "I think what we're saying is friendly to the consumer," Ciccone says. "If we're building the capacity, what we're doing is trying to defray some of the cost from consumers to the business end of this."

AT&T's critics don't buy this claim. They argue that by slicing up the Internet into different lanes, broadband companies are violating one of the basic network design principles responsible for the Internet's rise and amazing success. They add, too, that there's no proof that AT&T's plan would result in reduced broadband costs for home customers. Instead, consumers could lose out in a big way. If AT&T's plan comes to pass, the dynamic Internet, where innovation rules and where content companies rise and fall on their own merit, would shrivel. By exploiting the weaknesses in current laws, telecom firms would gain an extraordinarily lucrative stake in the new media universe. In the same way that a corporation like Clear Channel controls the radio airwaves, companies like AT&T could become kingmakers in the online world, granting priority to content from which they stand to profit most. Britney Spears, anyone?

To understand why critics worry about the future of the Internet in the absence of what they call network neutrality, it helps to look at the underlying philosophy of the ubiquitous network. Engineers are fond of describing the Internet as a "dumb network," a designation that's meant to be a compliment. Unlike other large communications systems—phone or cable networks—the Internet was designed without a specific application in mind. The engineers who built the network didn't really know what it would be used for, so they kept it profoundly simple, making sure that the network performed very few functions of its own. Where other networks use a kind of "intelligence" to define what is and what isn't allowed on a system, the various machines that make up the Internet don't usually examine or act upon data; they just push it where it needs to go.

The smallest meaningful bit of information on the Internet is called a "packet"; anything you send or receive on the network, from an e-mail to an iTunes song, is composed of many packets. On the Internet, all packets are equal. Any one packet hurtling over the pipe to my house is treated more or less the same way as any other packet, regardless of where it comes from or what kind of information—video, voice, or just text—it represents. If I were to download a large Microsoft Word e-mail attachment at the same time that I were to stream a funny clip from Salon's Video Dog, the Internet won't make any effort to give the video clip more space on my line than the document, even if I may want it to. If the connection is too slow to accommodate both files at the same time, my video might slow down and sputter as the Word file hogs up the line—to the network, bits are bits, and a video is no more important than a Word file.

The notion that the Internet shouldn't perform special functions on network data is known as the "end-to-end principle." The idea, first outlined by computer scientists Jerome Saltzer, David Clark, and David Reed in 1984, is widely seen as a key to the network's success. It is precisely because the Internet doesn't have any intelligence of its own that it's been so useful for so many different kinds of things; the network works consistently and evenly for everyone, and, therefore, everyone is free to add their own brand of intelligence to it.

Today's largest broadband firms, though, aren't accustomed to running dumb networks built on the end-to-end principle. AT&T ran the phone network at its own behest—and the company usually benefited from it. Historically, in the telecom industry, "there's been this instinct toward control," says Tim Wu, a law professor at Columbia and a co-author of *Who Controls the Internet?* At firms like AT&T and Verizon, both of which have roots in the monopolistic old AT&T, there's now an effort afoot to reengineer parts of the Internet by introducing more intelligence to manage and control data.

One firm that has been a vocal proponent of prioritizing data is Cisco, the giant network equipment company whose products currently power much of the Internet. "We think that as people use their broadband connections more intensively, the need to manage traffic is going to increase," says Jeff Campbell, director of government affairs at Cisco. The company has designed an array of products that allows service providers like AT&T and Verizon to scrutinize everything on their networks extremely closely. One Cisco brochure (PDF) touts a system called the Cisco Service Control Engine, which is described as "a deep packet inspection engine that helps enable service providers to identify, classify, monitor, and control traffic" on the network. "Deep packet inspection" refers to the practice of looking at each slice of data on the network and determining exactly what kind of information it is—whether it's part of an e-mail message, or a bit of a video file you're trading over Bittorrent, or perhaps a *New York Times* news story on the Web.

After examining each packet and deciding which user asked for it, where it's coming from, and what application it's meant for, the Cisco system allows network operators to assign various network privileges to the data. During a time of network congestion, data that is "delay sensitive"—like part of a voice phone call or a streaming video—can be moved along the network in a hurry, while packets that represent less urgent data—peer-to-peer file transfers or downloads of e-mail attachments—might be put on a slow lane. In this sort of network, were I to download a video file and a Word file at the same time, the network would notice it and might decide to slow down the Word file so that the video file would play smoothly.

Many Web entrepreneurs and network policy experts think that giving priority to some traffic is good for the Internet. In February 2006, Mark Cuban, the billionaire media entrepreneur and sports-team owner, posted a rant on his blog decrying the current state of network traffic management and calling on broadband firms to offer high-speed service for some kinds of data. "There are some basic facts about the Internet that remind me of driving on the 405 in Los Angeles," Cuban wrote. "Traffic jams happen. There is no end in sight for those traffic jams. The traffic jams are worse at certain times of the day. Whether it's the 405 or the

Internet." If we use car pool lanes to allow some cars to bypass traffic on our freeways, Cuban asked, why not add high-occupancy vehicle (HOV) lanes to the Internet, so that media that needs fast service can get to its destination more quickly?

Cuban is a cofounder of HDNet, a high-definition cable and satellite TV network, and has a particular interest in seeing the Internet give special treatment to certain files. In fact, the new Internet schemes are specifically designed to boost audio and video on the network. If your Word file slows down for a half second during download, you're not going to notice it; but if your Internet phone call has a half-second interruption, it would annoy you to no end.

Opponents of neutrality regulations say other applications currently being designed for the Internet will only work well if the network is improved. For instance, imagine if you were watching an Internet TV broadcast of a basketball game that allowed you to switch to different camera angles during the game. That program would be only useful, says Campbell of Cisco, if the camera angles appeared instantly, not seconds after you switched. Other advocates point to new medical diagnostic devices with which hospitals can monitor the status of patients at home; in that situation, it would seem obvious to give such traffic priority.

"I guess we could leave the Internet in the dark ages and leave everything as an unprioritized, unorganized mass where all bits are treated the same," says Campbell. "But we think good network management technology will improve overall performance, and consumers will have a better experience in the long term."

Despite Cisco's position, there is fractious division among network engineers on whether prioritizing certain time-sensitive traffic would actually improve network performance. Introducing intelligence into the Internet also introduces complexity, and that can reduce how well the network works. Indeed, one of the main reasons scientists first espoused the end-to-end principle is to make networks efficient; it seemed obvious that analyzing each packet that passes over the Internet would add some computational demands to the system.

Gary Bachula, vice president for external affairs of Internet2, a nonprofit project by universities and corporations to build an extremely fast and large network, argues that managing online traffic just doesn't work very well. At the February Senate hearing, he testified that when Internet2 began setting up its large network, called Abilene, "our engineers started with the assumption that we should find technical ways of prioritizing certain kinds of bits, such as streaming video, or video conferencing, in order to assure that they arrive without delay. As it developed, though, all of our research and practical experience supported the conclusion that it was far more cost effective to simply provide more bandwidth. With enough bandwidth in the network, there is no congestion, and video bits do not need preferential treatment."

Today, Bachula continued, "our Abilene network does not give preferential treatment to anyone's bits, but our users routinely experiment with streaming HDTV, hold thousands of high-quality two-way videoconferences simultaneously, and transfer huge files of scientific data around the globe without loss of packets."

Not only is adding intelligence to a network not very useful, Bachula pointed out, it's not very cheap. A system that splits data into various lanes of traffic requires expensive equipment, both within the network and at people's homes. Right now, broadband companies are spending a great deal on things like set-top boxes, phone routers, and other equipment for their advanced services. "Simple is cheaper," Bachula said. "Complex is costly"—a cost that may well be passed on to customers.

Expensive as they may be, the new network schemes will allow for myriad moneymaking opportunities. The new technology will allow AT&T and company to reserve the fast lane for the highest bidders. And AT&T says such a plan is perfectly fair. "It costs a lot to maintain and operate a network," says Ciccone of AT&T. "You don't pay for that by offering a raw pipe. We didn't build a copper line network 100 years ago so people could do whatever they want on it. We offered a phone service. And you don't build networks so that somebody else can necessarily use them for free. We have the capability through dedicated lines of service for offering a high-quality product. There's a service there. We should be able to offer that in the market."

Ciccone is particularly galled by the fact that those who are the most opposed to AT&T's plans are enormous firms—such as Google—that want to make money by offering video services online. "This really is just coming from a couple companies who have plans to stream movies," he says. "They hide behind the guise of the innovator in the garage who's building the next big Google. That's a lot of hooey because the little guy is not streaming movies. This is about the companies that want to stream movies, and they want to not just compete with us but with cable companies in doing so. What disturbs them is that we're building network capacity to be able to accommodate ourselves with a very high-quality product, and the Googles won't be able to deliver the same quality."

Technology companies do say they fear AT&T's network won't provide a level playing field and that AT&T's competitors won't be able to deliver videos that work as well as AT&T's content. Networks have finite space, and it is a fact of network engineering that when some data is given a priority on the network, other data will be pushed aside. At the Senate hearing, Stanford law professor and Internet policy expert Lawrence Lessig argued that this will put companies or individuals that can't pay for high-quality service at an enormous disadvantage, "reducing application or content competition on the Internet." In the past year, streaming-video Web sites have proliferated on the Internet, and some of the most popular services have come from start-ups like YouTube. Under AT&T's plan, flush firms like Google would be able to pay for all the space on the line, leaving the smaller guys out of luck. The Internet has long been a meritocracy, where smart and creative companies can act quickly and beat out established players. That wouldn't be so on AT&T's Internet.

Broadband operators respond by declaring they will offer high-speed services to all companies, big or small, and anybody will be able to pay for a spot in the fast lane. "Generally companies shy away from doing exclusive deals," says AT&T's Ciccone. "You don't say I'm only going to provide telephone service to only one bank." But as Amazon's Misener points out, "This is a zero-sum game. If you prioritize anyone's content you necessarily degrade someone else's. That's how it works." When you convert one lane on a freeway into a toll lane, it's true that you make traffic better for cars that can pay. But you also make traffic worse for cars that cannot.

Indeed, that's what makes AT&T's plan so lucrative. The company can't offer fast service to everyone. If it did offer all companies access to the fast lane for a low fee, the lane would soon become congested, and nobody would have an incentive to pay. To make the most money, the network operators may charge just a few firms huge sums to ride on the pipe. This means that one or two companies could lock in a preferred position on the network.

And AT&T's own services could benefit greatly from the new plan. For instance, AT&T offers a voice-over-the-Internet phone plan called CallVantage that competes with Skype, a free service owned by eBay. "Let's say there's a certain amount of revenue in voice services, maybe $125 billion in voice," explains Wu. If AT&T determines that letting Skype onto the fast lane will cause it to lose customers and, thus, revenue, it could decide to only let Skype ride the slow lanes. "If you're going to lose $10 billion to Skype by letting them on, why give them that money?" Wu says that under current regulations, this practice would be perfectly legal.

While such deals may be legal, AT&T says, they would be bad for business. If a broadband company didn't allow a popular service like Skype a spot in the fast lane, consumers would choose a different provider. "If you do make dumb decisions, your customers go somewhere else," Ciccone says. "Nobody wants to offer half a service with only special deals or arrangements for something of that nature. You're competing against other companies that may do it differently."

But if you don't like your Internet provider, would you really be able to go elsewhere? Cerf, who is now Google's chief Internet "evangelist," pointed out in the Senate hearing that only 53 percent of Americans now have a choice between cable modem and DSL high-speed Internet service at home. According to the FCC, 28 percent of Americans have only one of these options for broadband Internet access, and 19 percent have no option at all.

Moreover, phone and cable companies have been trying to reduce competition in the broadband business even further. They convinced the FCC to allow them to prohibit rival Internet service providers—such as Earthlink—from offering high-speed Net access on phone- and cable-company-owned lines. (Phone and cable companies do lease their lines to independent ISPs like Earthlink, but under current rules they can decline to do so at any time.) AT&T, Verizon, and Comcast have also pushed hard to stop cities across the country from launching free or low-cost municipal wireless Internet systems.

In this marketplace, if your DSL or cable modem provider begins to favor some content over others, you will have very little recourse. Even if you could choose another provider, doing so isn't easy. "It's not like there are two supermarkets in town and if you don't like one you can go to the other," Amazon's Misener says. He adds that "every economic theory we know suggests that when there's a duopoly"—in this case between cable broadband and phone broadband—"there will be tacit collusion in the market." So even if you could choose between broadband or cable service, eventually, as with radio stations in any metro area, you will find they all sound the same. Or think about your cable lineup. When your provider doesn't carry the TV network you like, what choice do you have? Almost none.

At the moment, there are very few regulations that outline what broadband companies can and cannot do with content on their lines. So far, the FCC has only been willing to outline some principles to which firms should adhere. In a speech in Boulder, Colorado, in February 2004, Michael Powell, the former FCC head, said that he didn't see the need for regulation. Instead, he set out a list of "Internet freedoms" that he "challenged the broadband network industry to preserve." Specifically, Powell called on high-speed network providers to allow their customers to access any legal content on the Internet, use any legal applications, and plug in any devices to their networks. The FCC later outlined these principles in a "policy statement" and imposed these conditions on Verizon and AT&T as temporary conditions of the mergers the companies underwent last year.

But while these "freedoms" allow customers access to any services, they don't outline whether AT&T can give some content priority on the network. In addition, there is a debate about whether Powell's "challenge" is enforceable at all. Last year, when one small North Carolina ISP began blocking Internet voice calls on its network, the FCC quickly stepped in and fined the firm. Telecom firms say the incident proves that the FCC has enough authority to block egregious behavior. But AT&T's Ciccone also acknowledges that adhering to the FCC's vision is a "voluntary commitment. It's not a rule or a regulation of the FCC. They laid out the broadband principles, and our compliance is purely a voluntary act on our part."

Wu explains the issue this way: "Right now it's like the ghost of Michael Powell has his finger in the dike" protecting us against the worst behavior of big companies. But if you were starting a new service on the Internet, "do you want to bet your business on the ghost of Michael Powell?"

Today, as numerous proposals for reforming telecom law float around Congress, broadband firms are fighting hard against a neutral network and apparently winning. (AT&T may certainly be on the government's good side, as it has been secretly allowing the National Security Agency to monitor its phone and Internet lines, according to a retired AT&T technician, as reported by Wired News.) In a party-line vote last week, Republicans on a House subcommittee defeated one neutrality proposal. According to many observers, another bill in the Senate, offered by Democratic Senator Ron Wyden of Oregon, faces similar dim prospects. In addition to lobbying, broadband firms have launched a campaign aimed at urging Americans to join their fight. Large telecom firms back a "coalition" called Hands off the Internet, which argues that instituting network neutrality amounts to government "regulation" of the Internet. On its

Web site, the group—which is funded by, among other companies, AT&T, and is headed by former Bill Clinton press secretary Mike McCurry—beseeches, "Join us and say NO to government regulation of the Internet!"

Opponents say that regulation is the only way to save the Internet from the likes of AT&T. "They would have the pipe split between the public Internet—which might get 1 Mbps speeds—and a toll lane on the rest of the 100 Mbps pipe they're laying," Tod Cohen, the director of government affairs at eBay, says of AT&T's plans. By "public Internet," Cohen is referring to today's Internet, the Internet of Google, Blogger, Skype, YouTube, and Flickr, services that came out of nowhere and are now indispensable. "They're saying, 'We'll leave the public Internet to be like the public-access station. But if you want to be on one of the fast channels, you have to pay.'"

Consumer advocate Chester sees a dark future for the Internet if big companies like AT&T gain unregulated control. "I think the public requires a serious national debate about what this means and what it's going to look like," he says. "There's a basic assumption that the Internet is going to remain forever open and diverse and affordable. I'm saying we should be cautious. We should really understand what these proposals mean for the kind of diverse voices we would want to see online."

Julian Dibbell

Dragon Slayers or Tax Evaders

*Buying and selling imaginary goods in computer
game worlds is big business. But it's left gamers
and IRS agents scratching their heads over the
real-world taxes owed on virtual treasures.*

If you haven't misspent hours battling an Arctic Ogre Lord
near an Ice Dungeon or been equally profligate spending
time reading the published works of the Internal Revenue
Service, you probably haven't wondered whether the U.S.
government will someday tax your virtual winnings from
games played over the Internet. The real question is, Why
hasn't it happened already?

Gamers who play EverQuest, an online game with
450,000 subscribers—playing parts that range from
Frogloks, a race of sentient amphibians, to Vah Shir, a regal
feline race—generate in their virtual world the kind of
imaginary economic activity that can be measured in real-
world terms like *gross domestic product*. According to Indiana University economist Edward Castronova, EverQuest's
annual GDP—the total wealth in goods and services an
economy creates—is about $135 million, or around half the
GDP of the Caribbean island nation of Dominica. Castronova sized up the virtual economy (in a widely read 2001

paper that launched his career as the Adam Smith of video games) in part to mock the pomposity of traditional econometrics. But the finding of the paper was no joke. EverQuest's subscribers are playing a game, to be sure, yet like all massively multiplayer online games, often called MMOs, it's one with remarkable economic potential.

To play an MMO is to engage in a quest. You take a character into a virtual world to hunt monsters, seek treasure, and enjoy the thrill of slowly rising from humble beginnings to imaginary wealth and stature. The rub, however, is that you can rise only so far without finding treasures placed in the virtual world by the game's creators or acquiring those items from other players. There are various ways to convince your playmates to surrender their weapons, magic spells, and mineral ores. Killing their characters is one way and befriending them another, but the quickest method is to offer fellow gamers real money. This is rarely approved by official rules of MMOs, but the practice is so widespread that if you look on online marketplaces like eBay today, you will find a thriving, multimillion-dollar market in Golden Runic Hammers, Ethereal Mounts, and similarly exotic items—all of them won (and anything found or wrested from another player is "won" in the context of a game) or bartered for in this or that MMO quest and many of them fetching prices in the hundreds, even thousands, of dollars.

In June 2003, I set myself the following challenge, posting it on my Web log for the world to see.

On April 15, 2004, I will truthfully report to the IRS that my primary source of income is the sale of imaginary goods—and that I earn more from it, on a monthly basis, than I have ever earned as a professional writer.

In the course of this project, I made a total of $11,000 selling on eBay the items I won playing a game called Ultima Online, $3,900 of which was in the final, most profitable month. I reported my profit to the IRS, and I paid the requisite taxes. But after I did so, a troublesome set of questions continued to nag at me—for which even IRS publication 525, entitled "Taxable and Nontaxable Income," couldn't provide answers.

This was remarkable, for publication 525 would appear to contain every conceivable form of income known to accounting. Here you'll find a description of gains, ill gotten and otherwise, so irregular that they can be taxed only according to that form of guesswork known as fair market value. Here are stocks, options, retirement watches, and stolen goods ("If you steal property, you must report its fair market value in your income in the year you steal it unless in the same year, you return it to its rightful owner").

Most significant for my purposes, here too are items acquired either through barter or as prizes in a game. The rules make clear the IRS's fundamental point: Goods taken in trade or won at play are taxable the moment they fall into somebody's hands, even if they are not sold for money. The more I read, the more I wondered whether reporting the amount I had brought home from selling virtual items on eBay was enough to satisfy the IRS.

What about the assets I bartered for or won in the game but never sold in the real world, the suits of armor stashed here and there with their easily established fair market value? What if I traded those assets for their value in Ultima Online's official currency, the Britannian gold piece, rather than for dollars? Wouldn't it be easy to establish their value in dollars nonetheless and, if I owed American taxes on the exchange, put a number on the deal that the IRS could grasp and love? And what about all the other MMO players out there—how long could the IRS be expected in good conscience to leave the resulting millions of dollars in wealth untouched?

You might think that I was letting my imagination run away with me—I certainly hoped I was. I thought that a glance at past IRS practices would assure me that the feds would never dream of taxing assets that had not been turned into money. I thought wrong.

The IRS has taxed barter transactions that are remarkably similar to the ones that online players engage in every day. In the late 1970s, for example, dozens of so-called barter clubs sprang up around the United States, said Deborah Schenk, a tax professor at New York University School of Law. The clubs put out directories in which members listed themselves as providing accounting, window washing, or other types of services. Any member could buy those services with "trade dollars," a virtual currency like Britannian gold pieces, and a member could earn trade dollars by offering his or her own services. By 1980 these clubs were handling an estimated $200 million worth of transactions every year, and the IRS took notice. In a 1980 ruling, the agency said that barter club transactions produced taxable income, even though no actual money changed hands. A 1982 law made enforcement of the ruling easier by requiring the clubs to provide the IRS with information about every transaction.

Swapping a financial audit for a dental checkup seems different from trading a Runic Hammer for an Ethereal Mount, if only because audits and checkups are real. But what does "real" mean for tax purposes? Richard Schmalbeck, a tax professor at Duke University School of Law, said the IRS determines that something has "real" value when a similar good or service trades on a market for actual money. Transactions involving items with real value are taxable. He acknowledged, though, that not every situation is clear.

Schmalbeck described the case of David Zarin, a gambler who spent close to a year playing craps in a casino, essentially without leaving the casino. Zarin lost $3.4 million worth of chips he acquired on house credit, and he converted few, if any, into actual money. Unable to collect, the casino wrote off all but $500,000 of his debt, arguably presenting Zarin with a gift of $2.9 million. Was it taxable, as the IRS claimed? Schmalbeck said that several courts reached different conclusions, unable to agree on whether the gambling debt or the gift (or, perhaps, the gambler's year in the casino) was real in any meaningful sense, even though the chips had fallen where they did and had a monetary value.

In any case, with virtual goods from Internet games being traded every day for actual money on eBay, wouldn't Schmalbeck's theory about similar goods trading on actual markets mean that trades occurring exclusively in a game are taxable? I set out for my local tax office in South Bend, Indiana, to find out.

Arriving near the end of the workday, I took a chair until my number was called and then approached the help desk, where a tax official named John Knight looked up at me with a mix of weariness and curiosity. I took a deep breath and proceeded to describe my business and the economy that sustained it. I cited publication 525. I inquired about barter income, specifically the difference, if any, between a painting for which the owner paid $6,000 bartered for half a year's rent and a virtual castle with a value of $1,200 established on eBay bartered for 10 million pieces of virtual gold. If John Knight's response was typical, the IRS hasn't done much thinking about the matter.

"OK, so I got a fake jewel that's worth 80 million points, gives me all kinds of invincibility," said Knight, striving doggedly to nail down what I was talking about. "But I got two of them or don't want to play [anymore]. And I can go on eBay and sell my jewel to some other character?"

"Uh, yeah," I confirmed.

Knight considered the facts and offered a nonbinding opinion: "That's so weird."

He ventured to say that it was doubtful the IRS would treat virtual items as cash equivalents anytime soon. Until the Britannian gold piece trades in international money markets, or until the value of a virtual amulet is as widely recognized as that of a beer, he suggested, "I don't think we're recognizing Dungeon and Dragon [*sic*] currency as legal tender."

Because he wasn't in a position to offer a final word, however, Knight gave me a number for the IRS's Business and Specialty Tax Line. "Specialty" sounded about right, so I called and told my story to a telereceptionist, who routed me to a small-business specialist, who passed me along to a barter-income specialist, who identified herself as "Mrs. Clardy, badge number 7500416," listened in silence to my query about virtual economics—and then put me on hold.

When Mrs. Clardy returned, she was a bureaucrat transformed. "We just had this little *discussion,*" she said, almost giggling. "And it sounds to us like [the online trades you've described] would be—yes—Internet barter." Here she paused, whether to catch her breath or to let the conclusion sink in, I couldn't tell. "However," she went on, "there are no regs, there is no code, there are no rulings, to rely upon. This is our *opinion.*"

Mrs. Clardy suggested I seek a more authoritative judgment. A "private letter ruling," she assured me, was the IRS's definitive opinion, in writing, on a particular taxpayer's situation. And a letter ruling in my case, she believed, would probably be the closest the IRS had ever come to an opinion on the status of virtual income.

"The ramifications are *enormous,*" Mrs. Clardy exhorted. "Break new ground!"

I intended to, but Mrs. Clardy had neglected to mention the $650 fee stipulated in the letter-ruling request instructions, or the tax lawyer I would have to hire to write the request with any effectiveness, or the six months I would have to wait for a final response. Even if none of these obstacles had stood in the way, I finally had to ask myself: Was this really the kind of ground I wanted to break? Considering what the IRS had done with barter clubs, it seemed prudent not to be the game player who officially invited the agency to visit the world of MMOs and gave the feds the opening to tax virtual income. That decision might force game companies, as John Knight had put it, "to start sending out 1099s every time somebody gets a gold coin or a bag of grapes or a shiny emerald" in a game's virtual word. It would certainly transform the thrill of the online quest into distress for the legions of players who couldn't afford to pay their new taxes and would likely doom my fellow Frogloks, Vah Shir, and other characters to appalling fates.

Adam L. Penenberg

Revenge of the Nerds

Arin Crumley and Susan Buice were just art squids with a handful of credit cards, a digital camera, and very patient parents. Now they have a (long) shot at the big time. How the digital wave gives power to the little people and reshapes the way movies reach the world.

Like many of his peers, 21-year-old Arin Crumley, a tall, Twizzler-thin videographer living in Brooklyn, New York, went trawling for a girlfriend on the Internet, blasting notes to more than a hundred likely prospects who had posted personals on *Time Out New York*'s Web site. Shortly afterward, Susan Buice, also young, a self-styled "artist in theory, waitress in practice," clicked open his e-mail: "What made you move to NY? Do you have any more pix? I think I might find you hot."

Unlike the others on Crumley's hit list, Buice decided to give him a chance and told him to drop by the restaurant where she worked the late-night shift. Crumley showed up but in disguise—sunglasses, a baseball cap—packing a video camera and snapping surreptitious candids and then trailing her as she left the restaurant for the subway. "Dear Stalker," Buice replied, after the photos arrived in her inbox. "So this is what the world sees. Just an innocent bystander. So pedestrian. Nothing like the tragic hero I feel as I trudge through each day." She told him the typical date wouldn't do justice to the stalking experience. "We need to think of another unique scenario—something challenging." He suggested they communicate without speaking, to avoid small talk.

For their first date, they wandered the Brooklyn waterfront, passing notes, drawing pictures, listening to music on each other's iPods, but not talking. Later, when Buice attended an artist colony in Vermont, they mailed videos back and forth; six months after meeting, they moved in together. Along the way, they amassed a collection of artifacts most couples would call "keepsakes." Buice and Crumley considered them artistic "by-products."

Eventually, in the way of youth the world over, they concluded that their courtship had to be immortalized—and that only a full-length feature film would suffice. They quit their jobs, pooled $10,000 in savings, lined up a stack of credit cards, and flew in a friend from the Left Coast to operate their prized possession: a Panasonic DVX-100 digital video camera. The saga of *Four Eyed Monsters,* their self-directed, self-obsessed movie, had begun.

It was an unlikely way to make a movie, and if it sounds self-indulgent and a tad "meta," well, it is. But we live in an age when the tools of self-expression have never been more accessible. Until recently, making a movie meant using a shaky Super 8 or low-resolution camcorder—or taking a flier that required tens of millions of dollars, hundreds of personnel, and superior technical expertise. It also meant dueling with the studio executives and distributors who decided which movies made it into theaters and which didn't—and who exerted ham-fisted control over the industry, making it all but impossible for neophytes such as Buice and Crumley to break through. (And even if they did, they were often roundly fleeced: bought off with a nominal take-it-or-leave-it offer, stripped of control of their work, and sent packing back to Mom.)

But the great digital push is under way. Now the same pair of lovelorn kids who would have vanished completely in another age can pick up a camera, teach themselves the art of filmmaking for next to nothing, and make a commercial-quality movie. They might even hit it big.

Of course, digital movies are not new. A decade ago, *Love God,* long since forgotten, was one of the first—if not *the* first—independent films entirely shot and edited in digital video. Back then, the format was merely a curiosity; now, with the price of a decent camera dipping below $4,000 and quality steadily improving, digital is reshaping entertainment as the talkies did 80 years ago, with similarly revolutionary effects. Even the notion of a "film" has begun to seem a little quaint: Sure, there are still your standard 90-odd-minute narratives, and they may be around forever, but because moving images are increasingly being viewed in and over a variety of venues and devices—from 3-D high-definition digital theaters to TVs to laptops to PDAs, cell phones, iPods, and everything between—even that form is morphing. A film today might be a series of 3-to 5-minute episodes or a 20-minute short. The Beastie Boys handed out 50 Hi8 videocams so that fans could capture them in concert (the band later returned them for a refund). For $164,000, South African director Aryan Kaganof created *SMS Sugar Man,* the first feature-length movie shot entirely on cell-phone cameras.

As a practical matter, digital opens up other opportunities as well. Not only is it cheap (35mm film costs about 200 times more than digital tape) but it's also lightweight, simple, and subtle. For documentary filmmakers, for example, the format lets them make movies they couldn't do otherwise. James Longley, director of *Iraq in Fragments,* rode in the back of a pickup truck filled with Mahdi Army militia, recording everything as they arrested alcohol sellers in a local market and later interrogated them. "A large part of being able to record that kind of material is the ability to be unobtrusive," Longley says, "to let the mechanism of the camera nearly vanish, unencumbered by lights, sound recordists, and film-changing bags."

With some 18 billion videos streamed online in 2005—up 50 percent from 2004—it's not surprising that new businesses are sprouting up around this digital explosion. Each day on YouTube, more than 40 million video views are delivered and 35,000 new clips are uploaded. Google and Yahoo have video search sites and large caches of moving content. Apple's iTunes Music Store sold 12 million video clips for $1.99 each over the span of just a few months. A new company out of Berkeley, California, called Dabble is vying to become a micro movie studio for the masses by inviting users to create, remix, browse, and organize video online. These aggregators are fast becoming the central nodes of an entirely new video marketing and distribution system, one far from Hollywood's control.

"Things that are authentic have great appeal," says Dabble founder and CEO Mary Hodder, who believes that movie studios will increasingly find themselves competing with films made by the masses. "Part of the issue is lowering the transaction cost for making and distributing programs and films, and part of it is the low cost of user-generated content, which is usually free. It's totally disintermediating."

That process of cutting out the middleman, while still in its infancy, has the potential to upend the balance of power that has governed the film industry for decades.

One of the early demonstrations that homespun digital could deliver Hollywood-worthy numbers was the 2003 release of *Open Water,* a psychothriller with a cast of two, a crew of three, and lots of sharks. Chris Kentis, who had been cutting film trailers for a production company, pounded out a script, auditioned actors, bought two digital video cameras, and shot the movie over the course of two years, mostly on the open ocean. "We were chasing weather, and weather was chasing us," Kentis says. "We could see a storm approaching with lightning, and the captain would tell us we had 15 minutes to shoot the scene. We could react immediately. It was guerrilla filmmaking on the water."

Kentis submitted the film to Sundance with low expectations, figuring that festival organizers would take a dim view of a movie with so few credits (written by Chris Kentis, directed by Chris Kentis, cinematography by Chris Kentis . . .). But not only was *Open Water* accepted, it attracted a distribution deal from Lion's Gate and, later that year, opened at 2,700 theaters across the country. Made for $130,000, it went on to rake in about $30 million at the domestic box office and close to $100 million worldwide (counting DVD sales).

At the outset, Buice and Crumley had none of Kentis's skill. For starters, they had no idea how to frame a shot or do basic cinematography. Their cameraman wasn't familiar with the camera. They had never acted before. After each shoot, they beamed the dailies on a wall of their cramped loft and edited footage on a Mac G5 computer with Final Cut Pro software. "Sometimes we ended up shooting and reshooting a scene four or five times before we got it right," Crumley says.

Then, a year into the project, with our young heroes maxed out on seven credit cards, *Four Eyed Monsters* was accepted into Slamdance, the rogue sidekick to the Sundance Film Festival in Park City, Utah. That led to invitations to other festivals—18 in all, including South by Southwest, the Sonoma Valley Film Festival, and Gen Art. Along the way, the pair collected several awards and glowing reviews. *Variety* called *Four Eyed Monsters* "fascinating," a film that "deliberately smudges the line between nonfiction and invention." The *Boston Phoenix* said it was "spry, brainy, endlessly inventive," an "*Annie Hall* of the 25-year-old set." Others described it as "exhilarating," "accomplished and endearing," a movie that contains "frantic vibrancy" and "delivers a powerful narrative punch." It felt like the beginning of a Nora Ephron script that would, inevitably, end with Buice and Crumley better dressed, in a fabulous apartment, and very much in love.

In fact, their whirlwind tour yielded nothing. After 18 festivals, they were still without a distribution deal to get *Four Eyed Monsters* into theaters. Their work seemed to resonate, but they had no money and no access to the pipeline. All they had was a mounting sense that people liked the thing. "We had a film that nobody knew about and nobody wanted to distribute," Crumley says. "Companies told us that the 'target' audience for our film was 'hard to pin down.' What they meant was that they had no tried-and-true formula for how to release a film to the type of audience our film appealed to, so they didn't want to take a risk."

Which got them thinking. At the time, social-networking site MySpace had about half the 75 million members it has today. But, children of the Web that they are, Buice and Crumley understood that social-networking sites could generate interest and create buzz—a free, self-fulfilling wave of publicity. A veritable army of users, the vast majority under 30, would market the film for them virally by blogging about it or posting video clips. As any ad person will tell you, that kind of lightning-fast word of mouth is the most powerful form of marketing, and the Internet makes it possible on a scale never before seen. So Buice and Crumley embraced what Crumley calls "collective curation"—the idea that a loyal, intimate, motivated fan base is better able to judge quality than any individual and that a thumbs-up from the "Netgeist" can be life changing.

It was here that Buice and Crumley began butting up against the film establishment. And it is here that the Web potentially becomes a transformative vehicle for independent filmmakers looking to crash the gates of the old system.

The pair's strategy began taking shape when they attended South by Southwest, where they posted a daily online video diary of their experiences. The diary proved to be popular and helped draw an audience to their screenings. For Buice and Crumley, it also drove home the point that the battle to get their film recognized (not to mention the peripheral tales of their own interpersonal combat and financial woe) was something their peers could relate to—that the story about their story might help them get over the hump.

So they sketched out a series of 10 three-to five-minute video podcasts that they intended to post once a month, a kind of *Project Greenlight* reality show about the making of their movie. One episode tells how they got started. Another relates their experience at Slamdance, detailing fights that broke out when an acting teacher and some of his students who appeared in the film clamored for more credit. A third looked at the impact the film was having on their relationship, which often seemed on the brink of disintegration. The first podcast was posted in November and immediately became a hit. It didn't take long for each new installment to attract 65,000 downloads via iTunes, YouTube, Google Video, MySpace, and other sites. As of late May, the first seven episodes had been downloaded about half a million times, unleashing a platoon of citizen marketers for the movie as the clips get posted on individual profiles, e-mailed down the line to friends, or played on iPods.

In fact, the marketing has been so effective that a recent *Four Eyed Monsters* screening at the Brooklyn Museum sold out completely: 470 tickets, in five minutes. In September, after Buice and Crumley post their 10th and final podcast, they plan to throw a series of screening parties across the country, organized by volunteers. They are trying out a new service from Withoutabox, which lets filmmakers distribute their work to art-house theaters nationwide; the more advance tickets sold, the more theaters sign on. Simultaneously, they'll release *Four Eyed Monsters* through their Web site, www.foureyedmonsters.com, either on DVD (complete with the podcasts, for around $15) or as a low-end digital download.

Crumley sees their strategy as a showdown between Hollywood and the New Economy: "If we are successful with releasing our film by building our own audience, getting the film directly to that audience, and using what they say about it to get to a bigger audience—that would prove that the distributors' existence is completely unnecessary. It's better than a couple of guys in an office making a multimillion-dollar decision based on their own personal taste."

Buice and Crumley are early examples of the millions who will flow into this filtration process. It's a process that may have some unlikely by-products of its own—possible alliances, say, between low-rent filmmakers and theater owners, who in the past have been separated by a gulf of power and influence. The rise of digital makes it easy enough to imagine a day when theater owners look to the Internet to find movies (or compilations of shorts, or animation, or any other sort of content that can be screened) to supplement the list coming out of Hollywood or the not-so-indies. Even now, they could gauge an audience's interest in advance—based on metrics such as the number of downloads or click-throughs—and deliver specialized, micro-targeted content on nights when they're tired of showing *Rocky VI* to an empty house. Remember, Buice and Crumley's online presence translated into a packed theater, with real tickets.

Of course, there are a number of things delaying that day. It would cost about $3 billion to convert all 36,000 movie screens in this country to digital, and the process has barely begun. But as Bud Mayo, president of AccessIT, a company that offers financing and expertise to theaters looking to convert to digital, explains, delaying that process leaves a lot of money on the table. "You have a $9 billion domestic box office, and that's using 15 percent of available seats," he says. "If you can impose digital cinema and all its benefits, and attract 5 percent more customers to fill some of those empty seats, that's a $3 billion to $4 billion opportunity."

Adding to the power of the digital model is the fact that the big studios themselves stand to benefit from digital conversion. The studios now spend upward of $10 million to dispatch 35mm prints of a single blockbuster, at $1,500 per print, to the thousands of theaters across the country (and $5 million for a more modest release). A digital release would bring that number down to about $200 per movie, transmitted at the push of a button.

Within 10 years, your local cinema will probably have digital capacity (although it may retain film technology for a time, to appease the purist Hollywood directors), and this will profoundly shift the economics of the movie business. AccessIT alone plans to convert 10,000 screens by 2010, by ponying up the up-front costs. (In return, film studios pay a $1,000 fee per screen the first time a film is shown, which Mayo says would generate about $15,000 a year per screen.) AccessIT then transports the film via satellite or fiber-optic cable from a central server; theater owners could add advertisements or trailers, track concessions, and even monitor lights remotely. They would have the technical capability to change their lineup on the fly, substituting 3-D rock concerts, video-game competitions, religious revivals, World Cup matches, versions of a movie dubbed in Chinese or Spanish, or indie efforts like *Four Eyed Monsters*.

"The studios are afraid of the loss of control that digital sets forth," says Ira Deutchman, president and CEO of Emerging Pictures, another company that converts theaters to digital. "Once you have digital equipment, exhibitors can play whatever they want on a given day. This changes the balance of power between exhibitors and Hollywood." Standing between the studios and the theaters, however, are the distributors, and they still wield enormous clout. "If a theater pulls a movie before the contract stipulates, it's going to have a problem," says David Zelon, head of production for Mandalay Filmed Entertainment. "It's worse than getting sued: You won't get the next big picture."

Zelon says Hollywood studios aren't losing any sleep over the ascendance of digital filmmaking or its ancillary benefits to small-scale indie directors. "Every once in a while, you'll get a *My Big Fat Greek Wedding* or *The Blair Witch Project*. But you really need a studio-marketing campaign. At the end of the day, you need stars, because the first question people ask about a film is, 'Who's in it?'" he says. "Without mass marketing, you won't capture the attention of a mass audience, and the Internet is not a viable way to attract a mass audience."

Famous last words. But even if Zelon is right, that doesn't make movies marketed and distributed over the Web negligible. If 85 percent of a cinema's seats lie fallow every year, 100 smaller movies that attracted a following would become a force, while 1,000 events—from films to sports events to rock concerts—could represent a revolution.

How long could theater owners turn their backs on that kind of upside? If they caved, how long could distributors afford to withhold their films, especially as more theaters joined the digital ranks? And perhaps most important, how long would studios back distributors in that battle if they could deliver their films directly to theaters in a few minutes, for a fraction of the cost?

In the end, of course, Hollywood might respond to the threat posed by Buice, Crumley, and the rest by doing what it did to indie film in the 1990s: co-opting it. The Web could simply become a farm team, a place to scout talent. It seems a safe bet that someone at News Corporation, which paid more than half a billion dollars for MySpace, is watching the site closely for the next breakout star, whether of film, music, or anything else. (Even this scenario should give distributors the shakes, however.)

Meanwhile, filmmakers like Buice and Crumley are on their own, their Nora Ephron denouement deferred. Their 450-square-foot loft in a former factory building in Bushwick, Brooklyn, is awash in digital videotapes. An archaic gas stove is set right inside the front door; their bedroom is a mattress stuffed into an alcove, secreted behind a red curtain. A whiteboard is covered in notes, an almost endless to-do list. Two Apple monitors bathe the room in light.

The pair have racked up $54,000 in credit-card debt and are so broke that tonight's dinner consists of almond-butter sandwiches, which they'll eat on their way to the Apple Store in Soho, where they are scheduled to give a lecture. Still, asked if they would accept a $2 million offer from a distributor for the rights to *Four Eyed Monsters,* Crumley says, "No." Buice isn't so sure. "Only if we maintain control," Crumley insists. What if that wasn't part of the deal? "No."

Buice bites her tongue. She and Crumley fought almost every day while they were making their picture. She once got so fed up, she told him she was leaving as soon as they finished the damn thing. But their arguments smacked of truth. Authenticity, even. And in the end, they made the couple's story—what's the word?—cinematic.

Katharine Mieszkowski

"I Make $1.45 a Week and I Love It"

On Amazon Mechanical Turk, thousands
of people are happily being paid pennies to do
mind-numbing work. Is it a boon for the
bored or a virtual sweatshop?

A picture of a woman's pink shoe floats on my computer
screen. It's a flat, a street version of a ballet shoe. My job is to
categorize the shoe based on a list of basic colors: Is it red,
blue, pink, purple, white, green, yellow, multicolored? A
description next to it reads "Pink Lemonade Leather." This
is not exactly a brain-busting task; I'm doing it while talking
to a friend on the phone. With the mouse, I check a box
marked "pink." In the next split second, a picture of a navy
blue shirt appears. I check "blue." Assuming my answers
jibe with those of at least two other people being paid to
scrutinize the same pictures, I've just earned 4¢.

With my computer and Internet connection, I have
become part of a new global workforce, one of the thousands of anonymous human hands pulling the strings inside
of a Web site called Amazon Mechanical Turk. By color-
coding the clothing sold by the online retailer, which helps
customers to search for, well, pink shoes, I can now call
myself a Mechanical Turker. In this new virtual workplace,

everything is on a need-to-know basis, including who is doing the work; what the point of the work is; and, in some cases, the very identity of the company soliciting the work.

Launched in November 2005, Amazon Mechanical Turk is named after a legendary automaton from the 18th century, "the Turk," which could play chess. The wooden man, adorned with a turban, appeared to be powered by the machinery of a clock. He even checkmated Benjamin Franklin, a devotee of the game. The Turk was a sensation: a machine that seemed to think. Coinciding with the birth of the Industrial Revolution, the Turk heightened anxieties that machines would replace humans in the workplace. Of course, it turned out to be a fabulous hoax. The ghost in the machine was, as skeptics had suspected, all too human. A chess expert was hidden in the Turk, making all the right moves.

The 21st-century twist on the Turk, conceived by Amazon CEO Jeff Bezos, doesn't try to hide the people inside the machine. On the contrary, it celebrates the fact that we have become part of the machine. For fees ranging from dollars to single pennies per task, workers, who cheekily call themselves "turkers," do tasks that may be rote, like matching a color to a photograph, but can confound a computer. Conceived to help Amazon improve its own sites, Mturk.com is now a marketplace where many companies have solicited workers to do everything from transcribing podcasts for 19¢ a minute to writing blog posts for 50¢. Amazon takes a cut from every task performed.

Amazon claims its virtual workplace provides "artificial artificial intelligence"—a catchy way of saying human thought. "From a philosophical perspective, it's really turning the traditional computing paradigm on its head," says Adam Selipsky, vice president of product management and developer relations for Amazon Web Services. "Usually people get help from computers to do tasks. In this case, it is computers getting help from people to do tasks." As Tim O'Reilly, a computer book publisher and tech industry figure, puts it on his blog, old dreams of artificial intelligence are "being replaced by this new model, in which we are creating more intelligent systems by using humans as components of the application."

So who wants to be the human component of a computer application? A lot of people, it turns out. Since last November, thousands of workers from the United States and more than a hundred other countries have performed tasks on Mturk.com. The most dedicated turkers have even formed their own online communities, such as Turker Nation.

The companies are certainly happy. The ones I contacted remarked how stunningly little it costs them to get work done through Amazon Mechanical Turk. Divvying up projects to hundreds of people not only gets the job done more quickly than contracting it out to temps or consultants—much less an actual employee—it gets it done much more cheaply. The tag line for the site could be: dirt-cheap artificial artificial intelligence. One tech company that uses Mturk.com to answer troubleshooting questions brags that it pays tens of dollars on Mechanical Turk for work that would typically cost thousands. And, hey, if companies don't like the quality of the work they get from turkers, they simply don't pay them.

As soon as it launched, the Mechanical Turk site sparked a hue and cry in the blogosphere. "Amazon, you cheap bastard. Don't you at least have the decency to pay minimum wage?" demanded one poster on a tech site. Another commentator sneered that it peddled "jobs even illegal aliens won't do." There is something a little disturbing about a billionaire like Bezos dreaming up new ways to get ordinary folk to do work for him for pennies. Is a cut-rate pittance the logical result of tapping into a global workforce of people with a computer, an Internet connection, and an Amazon account? And, really, who are all these people working for a measly 1¢?

The real human ingenuity of Mechanical Turk shines in the novel ways that companies and workers use it to get tasks done efficiently. CastingWords is a transcription service built entirely on the work of Mechanical Turk transcribers and editors. With a little code, plus the turkers, it has succeeded in basically automating the process. The company charges its customers from 42¢ a minute for podcast transcription to 75¢ a minute for other audio. CastingWords pays Mechanical Turk workers as little as 19¢ a minute for transcription. If a transcription job is posted on Mechanical Turk for a couple of hours at the rate of 19¢ a minute, and no worker has taken on the project, the software simply assumes the price is too low and starts raising it.

After a transcription assignment is accepted by a worker, and completed, it goes back out on Mturk.com for quality assurance, where another worker is paid a few cents to verify that it's a faithful transcript of the audio. Then, the transcript goes back on Mturk.com a third time for editing and even a fourth time for a quality assurance check. "It's been terribly useful for us," says Nathan McFarland of Seattle, one of the cofounders of CastingWords. Transcription is the type of relatively steady task that keeps turkers with good ears who are fast typists coming back. "There are people who have been with us for months, and they're not leaving," says McFarland.

One of those workers is Kristy Milland, 27, a mother of one who runs an at-home day care in Toronto, as well as a Web site called RealityBBQ about the reality TV show *Big Brother*. "I have a lot of free time basically sitting at the computer while the kids play," she says. Among the work she does is editing and quality assurance for CastingWords, but not transcription, because she has tendinitis. When Mturk.com first began, Milland would churn through 3¢ HITs. (That's "human intelligence tasks," turker lingo for jobs.) Amazon was paying turkers to make sure that photos of businesses used on its A9 site, a local search engine, matched the actual businesses listed, a task a computer can't do. In an eight-hour day, when she didn't have the kids to watch, Milland could go through 1,000 photos, making a cool $30.

Lately, she's found a way to goose her earnings by competing for bonuses. A number of service companies use Mechanical Turk to do a "human augmented search." Say you're in a sports bar and having an argument about whether Roger Clemens has ever thrown a no-hitter. You can end the debate once and for all with a call to one of the services, which instantly posts the question on Mechanical Turk. Turkers then surf the Web and generally earn 2¢ for each answer.

Back in the sports bar, when you get the answer— "Clemens has never pitched a no-hitter"—you can rate the answer as great, good, lame, or junk. Answers deemed "lame" or "junk" are rejected, and the worker is not paid. If you don't rate the answer at all, the worker is automatically paid his or her 2¢ after seven days. Turkers who get the most "great" votes in a week get bonuses of as much as $75. In a good week, Milland can answer 100 research questions, making all of $2, but scoring one of those lucrative performance bonuses, she says, makes the search worthwhile.

The trivia pursuits are so competitive that they're snatched up by turkers within a minute of being posted. So Milland has set up software to notify her whenever a new question shows up on Mechanical Turk, so she can be the first to grab it. Plus, she's armed her Web browser with links to her top 100 reference sites so she can answer the questions as efficiently and accurately as possible. Turkers can choose to be paid in Amazon credit, making it easy to shop at the company store. Just the other day, Milland ordered $600 worth of DVDs and books for her family, as well as prizes for contests on her RealityBBQ. "It still doesn't add up to a lot of money per hour, but if I'm sitting there watching TV anyway, it's more than I'd make just sitting there," she says.

Milland's main beef with Mturk.com is that there's no way to complain if a company rips her off by refusing to pay for good, accurate work. "Amazon basically says, tough, they can reject what they want," she says. "There's no recourse." (Word of bad-apple companies, however, spreads fast on turker forums.) Milland would also like to see more work and more companies on Mturk.com. When the site first launched there was more to do, she says. These days it feels as if there are fewer opportunities and too many workers competing for them.

Of course, for all its rhetoric about artificial intelligence, Amazon did not launch Mechanical Turk for the good of science. For every task a worker completes for another company, the retailer collects a 10 percent fee from that company. For cheap HITs that pay just a penny, Amazon charges the company half a cent per HIT. Companies need not know the real name, much less the address or Social Security number, of turkers. Unless a worker earns more than $600 from a given company, the business has no obligation to issue the worker a tax form or report the earnings to the Internal Revenue Service. Few workers cross that $600 threshold with any one company. Yet workers are required to report the money they earn on Mturk.com to the IRS as income—yes, even the $1.45 I made—to be taxed at the high rates of the self-employed. There's no chance that a worker might land a full-time job with a company through Mechanical Turk, since it's expressly forbidden in the site's "participation agreement," which requires workers to submit all work through the site and not directly to the requester.

To a labor activist like Marcus Courtney of WashTech, a tech workers' union, the whole arrangement represents a dystopian vision of a virtual sweatshop. "What Amazon is trying to do is create the virtual day laborer hiring hall on the global scale to bid down wage rates to the advantage of the employer," he says. "Here you have a major global corporation, based in the United States, that's showing the dark side of globalization. If this is Jeff Bezos's vision of the future of work, I think that's a pretty scary vision, and we should be paying attention to that."

Rebecca Smith, a lawyer for the National Employment Law Project, seconds that. "The creativity of business in avoiding its responsibility to workers never ceases to astound," she says dryly. "It's day labor in the virtual world." Smith sees Mechanical Turk as just another scheme by companies to classify workers as independent contractors to avoid paying them minimum wage and overtime, complying with nondiscrimination laws, and being forced to carry unemployment insurance and workers' compensation. "It's an example of cyberspace overtaking a country's labor laws," she says. Needless to say, the turkers don't see it that way.

Curtis Taylor, 50, a corporate trainer in Clarksville, Indiana, who has earned more than $345 on Mturk.com, doesn't even think of turking as work. To him, it's a way to kill time. "I'm not in it to make money; I'm in it to goof off," he says. Taylor travels a lot for business and finds himself sitting around in hotel rooms at night. He doesn't like to watch TV much and says that turking beats playing free online poker. To him, it's "mad money," which he blows buying gifts on Amazon, like Bill Bennett's *America, the Last Best Hope,* for his son, a junior in high school. "If I ever stop being entertained, I'll stop doing it," he says. "I'll just quit."

Yet what's a happy diversion for Taylor is serious business for the companies on Amazon Mechanical Turk. Efficient Frontier, a search engine marketing firm, has used Mturk.com to accomplish tens of thousands of tasks since early 2006. Efficient Frontier helps companies figure out which keywords will bring Web surfers to their sites. With Mturk.com, Efficient Frontier can afford to pay three different people to look at each potential keyword and vote whether those words are relevant to a given site. It costs the company just 4.5¢ to test each keyword, paying 1.5¢ to Amazon and 1¢ each to three turkers.

Sherwood Stranieri of SkyPromote, another search engine marketing firm in Boston, says he now has a virtual staff of 120 workers on Mturk.com. "It's like a giant human computer," he says. "It's like having an infinite attention span." Stranieri pays qualified turkers to surf a Web directory and figure out exactly where a specific site should be listed. He can get 300 of these tasks done in just five or six hours, even if he posts them on Mturk.com in the middle of the night. He pays 5¢ a task. "Pricing is very low right now because there are so many more workers than tasks right now," he says. "People are fishing around for work to do." Why do people do it if the pay is so low? It's a question Stranieri wonders about himself. "I think it's something of a hybrid between trying to make money on the side and a diversion, a substitute for doing a crossword puzzle. It's sort of a mental exercise."

Eric Cranston, 18, who recently graduated from high school, got into turking because at the time he didn't have anything better to do. "When it came out," he says, "I had just broken my foot, so I was just at home doing nothing on the computer. So, why not?" He's used the money he's made answering survey questions and transcribing podcasts to buy a game controller and a computer monitor. He recently transferred $200 to his bank account. "I don't think anyone could actually make a living off of Mturk. There isn't enough work," Cranston says. He is one turker, however, who is plotting how to move up the food chain. Currently, Cranston and a friend are working to launch a Web-based business altering photographs, called Image Den, using, naturally, Mechanical Turkers to treat the images.

In its earliest days, someone posted a request on Amazon Mechanical Turk, offering to pay 2¢ for a drawing of a sheep facing left. Peter Cohen, director of Amazon Mechanical Turk, says the company was "puzzled by" the request. The requester was Aaron Koblin, a student in UCLA's Design/Media Arts program, who was writing his master's thesis about the site. He was intrigued by Amazon's effort to "establish a framework for the utilization of people as computers," as he wrote in his thesis. "My project was very tongue in cheek," he tells me. "On the one hand, it's using the system the way it's meant to be used. On the other hand, it's asking them to do this ridiculous thing."

The grad student invited turkers to draw up to five sheep at the rate of 2¢ apiece. Over 40 days and 40 nights, the sheep flooded in at a rate of 11 per hour. By the end, 7,599 turkers had participated. He collected 12,000 sheep and promptly put 10,000 of them up for sale at the rate of $20 for 20 sheep at the Sheep Market. This caused some consternation among the people who had drawn them. "They're selling our sheep!!!" wailed one poster on a turker message board. Another wrote, "Does anyone remember signing over the rights to the drawings?" In fact, they had. To participate in Amazon Mechanical Turk, workers, in the legalese of the site, "agree that the work product of any Services you perform is deemed a 'work made for hire' for the benefit of the Requester, and all ownership rights, including worldwide intellectual property rights, will vest with the Requester immediately upon your performance of the Service."

Why sheep? Koblin relished all the associations that sheep have, from the biblical followers of the Good Shepherd to George Orwell's *Animal Farm*. The term *cottage industry* comes from peasants' setting up shop at home, when it wasn't planting or harvesting season, often spinning wool. "The cottage industry, which would employ entire families from their houses, has notable similarities to Mechanical Turk, such as employing people for spare time, working from home, and relative anonymity," he wrote in his thesis.

Koblin wanted his project to capture the creativity expressed by turkers, while drawing attention to the insignificant role each of them played in the process. He certainly succeeded in capturing their creativity. Even after he stopped accepting sheep and started selling them by the lot on the Internet, more people wrote to him wanting to contribute sheep for free. They just wanted to see their sheep join the herd. "Most of these people clearly weren't in it for the money," Koblin says. "They weren't doing it so they could get 2¢. It was more about participating in something larger."

Maybe so. But maybe the ultimate message is: Congrats, fellow humans, we're not obsolete! The machines, they still need us! Only at a very sheepish price.

John Gruber

Good Journalism

How to read tech news

There was a great story yesterday by "technology writer" Dan Goodin at the Associated Press, and because it was from the AP, we can read it on several sites:

- Washington Post: "Macs Are Virus Targets, Some Experts Warn"
- CNN: "Viruses Catch up to the Mac: Experts Debate Just How Susceptible Apple Is Becoming"
- MSNBC: "Macs No Longer Immune to Viruses, Experts Say: Apple's Growing Market Share, New Chips Said Making It More of a Target"
- Wired News: "Macs Invulnerable No More"
- Etc.

What's great about an AP story like this one is that if you're only paying attention to the headlines, it creates the impression that there are multiple reports from all over the Web corroborating the same point, when in fact it's just one story, repeated many times over by news publications that regurgitate whatever comes in over the AP wire.

Oh, and I love the way both the CNN and MSNBC subheads conflate the Mac—which is a computer that can in fact be attacked by computer viruses—with "Apple," which is a company and therefore, one would think, not possibly "susceptible" to or "targeted" by viruses. That's good journalism, as it keeps the readers on their toes. *Good god, now they're going after entire companies.*

Journalism this good deserves a close analysis. Let's start with the lead.

> Benjamin Daines was browsing the Web when he clicked on a series of links that promised pictures of an unreleased update to his computer's operating system.
>
> Instead, a window opened on the screen and strange commands ran as if the machine was under the control of someone—or something—else.

"Or something"—could it be gremlins? Or worse, poltergeists? Spooky.

> Daines was the victim of a computer virus.

Damn, no poltergeists. But Goodin had me there for a second.

So what we've got is the classic "trend piece opening with a vignette" *Mad Lib* formula.

> Such headaches are hardly unusual on PCs running Microsoft Corp.'s Windows operating system. Daines, however, was using a Mac—an Apple Computer Inc. machine often touted as being immune to such risks.

Oh, zing! Not just an opening vignette, but the *opening vignette with the ironic twist!* See, I thought he was going to say Daines was using a Windows PC, because, you know, that sort of scene plays out *millions of times a year* to Windows PC users. You'd think I would have seen the "but he was using a Mac" twist coming, given the various "Macs are not immune to viruses" headlines that the piece ran under. But Goodin's masterful storytelling bamboozled me.

Who exactly is touting the Mac as "immune to such risks"? Goodin doesn't say, but his word is good enough for me. I'm sure whoever they are, they're experts.

I, on the other hand, had never been under the impression that the Mac was either magically or technically "immune" or "invulnerable" to viruses, Trojan horses, spyware, adware, malware, and so forth. Rather, I thought it was simply the case that, for whatever reasons, such software isn't a problem for Mac users and hasn't been for the last 15 years or so. I.e., that Macs aren't magically protected, and that in theory, malware could be written to target the Mac, but that the point is that in practice, in the real world, they aren't.

On the other hand, Macs *do* happen to be immune to *Windows* viruses and spyware and adware and Trojan horses, thousands of which are discovered every month. But why sweat the details?

> He and at least one other person who clicked on the links were infected by what security experts call the first virus for Mac OS X, the operating system that has shipped with every Mac sold since 2001 and has survived virtually unscathed from the onslaught of malware unleashed on the Internet in recent years.

Good lord. Daines and "at least one other person"—that means at least two Mac users were hit by this virus. (And what's funny about the "onslaught of malware unleashed on the Internet" is that it isn't just Mac users who've emerged "virtually unscathed" but just about everyone who isn't using Microsoft Windows. It's enough to make you think that the problem isn't "Internet malware" but "Windows malware.")

What virus was it that hit Daines? What kind of havoc did it wreak? Goodin doesn't say—who cares about the details, really?—but judging from his description that Daines caught the virus after he "clicked on a series of links that promised pictures of an unreleased update to his computer's operating system," it seems safe to assume it was the *Oompa-Loompa* Trojan horse described by Ambrosia Software's Andrew Welch back in February.

And what does Oompa-Loompa do? It attempts to spread itself via iChat on Bonjour—but it can only be spread if the person whose Mac it attempts to infect opens its file attachment payload. So you can only get it via other Macs on your local network, and if you do, you have to manually open a file named "latestpics.tgz," and if you do, all it will do is attempt to send itself to other local Macs on iChat.

Devastating and unstoppable.

"It just shows people that no matter what kind of computer you use you are still open to some level of attack," said Daines, a 29-year-old British chemical engineer who once considered Macs invulnerable to such attacks.

Daines's uninformed opinion that Macs were "invulnerable" to such attacks indicates that he believed it was safe to download and open any random file from the Internet, including gzipped archives from the sort of sites that traffic in bootleg screen shots of future Mac OS X releases.

Perhaps someday I'll have an opportunity to make equally informed statements regarding chemical engineering—a subject about which I am utterly ignorant—in an Associated Press report.

> Apple's iconic status, growing market share and adoption of same microprocessors used in machines running Windows are making Macs a bigger target, some experts warn.

That's quite an interesting theory—that the malware plaguing so many millions of PCs running Windows isn't necessarily the result of problems with Windows itself but is rather the result of something related to their Intel "microprocessors."

I'll bet that means all the other operating systems that run on Intel-compatible ×86 processors, such as Linux and FreeBSD, are just as susceptible to malware as Windows.

> Apple's most recent wake-up call came last week, as a Southern California researcher reported seven new vulnerabilities. Tom Ferris said malicious Web sites can exploit the holes without a user's knowledge, potentially allowing a criminal to execute code remotely and gain access to passwords and other sensitive information.
>
> Ferris said he warned Apple of the vulnerabilities in January and February and that the company has yet to patch the holes, prompting him to compare the Cupertino-based computer maker to Microsoft three years ago, when the world's largest software company was criticized for being slow to respond to weaknesses in its products.

This is in contrast to Microsoft now, in 2006, when their Windows users are no longer plagued by security problems.

And we know for a fact that at least *two* Mac OS X users have been hit by Oompa-Loompa just this year.

The bugs reported by Ferris are legitimate bugs, but to my eyes (and to Rosyna Keller at Unsanity—who thinks Ferris is counting the same tagged image file format [TIFF] rendering bug twice), they're all just ways to make an application crash, one of which has already been fixed in 10.4.6. But Ferris reports that this one, regarding Safari, "causes the application to crash, and or [*sic*] may allow for an attacker to execute arbitrary code." Emphasis on the *may* in "may allow," apparently, because the only thing his examples do is cause Safari to crash.

Anything that causes Safari to crash certainly sucks. And presumably Apple is working not just to fix these particular bugs but to fix the architecture of Safari to make it less vulnerable in general to these sorts of bugs in the system's image-parsing routines. But the genius here—and I'm not sure whether the credit goes to Ferris or Goodin, so let's just credit them both—is in the leap from bugs that, as Ferris originally described, "may allow for an attacker to execute arbitrary code" to bugs that, in Goodin's article, "potentially [allow] a criminal to execute code remotely and gain access to passwords and other sensitive information."

Because, see, in Ferris's original report, he meant "may" in the sense that they may, or they may not, but that he didn't actually know whether it was possible and has no evidence that they could. But in Goodin's AP story, that changes to "potentially," which *means* "capable of being but not yet in existence; latent," which is good journalism because "potentially allowing a criminal to execute code remotely" is much scarier sounding than "definitely allowing a jerk to crash your Web browser."

"[Microsoft] didn't know how to deal with security,

and I think Apple is in the same situation now," said Ferris, himself a Mac user.

By "same situation," Ferris is referring to what? The two guys who were hit by the Oompa-Loompa Trojan horse? One can only hope that Apple will one day handle security issues as well as Microsoft does now.

Apple officials point to the company's virtually untarnished security track record and disputed claims that Mac OS X is more susceptible to attack now than in the past.
Apple plans to patch the holes reported by Ferris in the next automatic update of Mac OS X, and there have been no reports of them being exploited, spokeswoman Natalie Kerris said. She disagreed that the vulnerabilities make it possible for a criminal to run code on a targeted machine.

Classic he-said/she-said situation: He said criminals can take over your Mac; she said they can't. One way to resolve this would be to emphasize the fact that Ferris has no proof to back up his claim. But a good journalist like Goodin knows that would just take the oomph right out of the story. And oomph—not facts or accuracy—is at the heart of every good story.

In Daines' infection, a bug in the virus' code prevented it from doing much damage. Still, several of his operating system files were deleted, several new files were created and several applications, including a program for recording audio, were crippled.
Behind the scenes, the virus also managed to hijack his instant messaging program, so the rogue file was blasted to 10 people on his buddy list.

Blasted is a great word. Much more exciting than something more accurate, such as "sent as an attachment."

"A lot of Mac users are in denial and have blinders on that say, 'Nothing is ever going to get to us,'" said Neil Fryer, a computer security consultant who works for an international financial institution in Britain. "I can't say I agree with them."

Fryer, also a Mac user, said he has begun taking additional precautions over the past year to make sure he doesn't fall victim to an attack. He spends more time than in the past scrutinizing his security logs for signs of intruders, and he uses a firewall and additional security applications, just as he would with a Windows-based machine.

It's so obvious that horrible things would happen to Fryer's Mac if he hadn't taken these steps that there's no need to mention what those horrible things are. Next thing you know, Apple is going to have to start shipping a firewall as part of the OS. I can see it now: right there as a tab in the Sharing panel in System Preferences.

The Mac's vulnerability could also increase as Apple transitions to a product line that uses microprocessors made by Intel Corp., security experts said.

With new Macs running the same processor that powers Windows-based machines, far more people will know how to exploit weaknesses in Apple machines than in the past, when they ran on the PowerPC chips made by IBM Corp. and Motorola Corp. spinoff Freescale Semiconductor Inc.

"They have eliminated their genetic diversity," said independent security consultant Rodney Thayer. "The fear is that we're going to run into a new class of attacks."

Thayer's photograph accompanied the article in many publications. You can tell he's a genuine computer security expert because he has long black hair and a beard.

The article closes.

> But as Daines can attest, there are no guarantees.
> "We're all sort of waiting with bated breath to see if any problem will happen and the jury is still out," said Thayer, the independent security consultant. "I don't think you'll find a consensus."

Only here, at the very end, does Goodin's article fall short. Rather than closing with a devastating gut punch, it just sort of fades out with a reasonable "we'll see what happens" whimper.

With Apple yesterday launching a new television ad campaign that draws specific, pointed attention to the fact that Macs are not besieged by malware, this sort of bogus "trend piece" that purposefully conflates the issue of whether Macs are *in theory* potentially vulnerable to malware (yes) with whether Macs are *in reality* under attack (no) is just what the doctor ordered. But it deserves to end with another sucker punch.

If Goodin wanted to be reasonable or accurate, he could have written a story titled "Some Guy Double-Clicked a Trojan Horse Virus for Mac OS X but It Didn't Actually Spread to Anyone Else," but what kind of story would that be? OK, it'd be a true story, but it wouldn't be a good story.

No one would have linked to such a story except to make fun of it: What would be the point of making a big stink out of one guy who got hit by a Mac OS X Trojan horse—which was so poorly written that it couldn't even successfully spread to another computer—when there are hundreds of thousands (millions?) of Windows users suffering from malware every single day?

What good journalism calls for is taking that one guy and writing an article that presents his episode as though it were part of a trend of increasing Mac virus attacks. No one is going to make fun of Dan Goodin—or the Associated Press, or the dozens of reputable news outlets that ran the story—for that.

Paul Boutin

You Are What You Search

*AOL's data leak reveals the seven ways people
search the Web.*

AOL researchers recently published the search logs of
about 650,000 members—a total of 36,389,629 individual
searches. AOL's search nerds intended the files to be an academic resource but didn't consider that users might be
peeved to see their private queries become a research tool.
Last weekend, the Internet service provider tried to pull
back the data, but by that point it had leaked all over the
Web. If you've ever wanted to see what other people type
into search boxes, now's your chance.

The search records don't include users' names, but each
search is tagged with a number that's tied to a specific AOL
account. The *New York Times* quickly sussed out that AOL
Searcher No. 4417749 was 62-year-old Thelma Arnold.
Indeed, Arnold has a "dog who urinate on everything," just
as she'd typed into the search box. Valleywag has become
one of many clearinghouses for funny, bizarre, and painful
user profiles. The searches of AOL user No. 672368, for
example, morphed over several weeks from "you're pregnant he doesn't want the baby" to "foods to eat when preg-

nant" to "abortion clinics charlotte nc" to "can christians be forgiven for abortion."

While these case studies are good voyeuristic fodder, snooping through one user's life barely scratches the surface of this data trove. The startup company I work for, Splunk, makes software to search computer-generated log files. AOL's 36 million log entries might look like an Orwellian nightmare to you, but for us it's a user transaction case study to die for. Using the third-party site splunkd.com, I've parsed the AOL data to create a typology of AOL Search users. Which of the seven types of searcher are you?

THE PORNHOUND

Big surprise, there are millions of searches for mind-bendingly kinky stuff. User No. 927 is already an Internet legend—search for 927 from splunkd.com if you're not faint of heart (and not at the office). When I clicked Splunk's "Show Events by Time" button, though, I found that porn searchers vary not only by what they search for but when they search for it. Some users are on a quest for pornography at all hours, seeking little else from AOL. Another subgroup, including No. 927, search only within reliable time slots. The data doesn't list each user's time zone, but 11 p.m. Eastern and 11 p.m. Pacific appear to be prime time for porn on AOL's servers. My favorite plots show hours of G-rated searches before the user switches gears—what I call the Avenue Q Theory of Internet usage. User No. 190827 goes from "talking parrots jokes" and "poems about a red rose" before midnight to multiple clicks for "sexy dogs and hot girls" a half hour later. An important related discovery: Nobody knows how to spell *bestiality*.

THE MANHUNTER

The person who searches for other people. Again, I used Splunk's "Show Events by Time" function to plot name searches by date and time. Surprisingly, I didn't uncover many long-term stalkers. Most of the data showed bursts of searches for a specific name only once, all within an hour or a day, and then never again. Maybe these folks are background checking job candidates, maybe they're looking up the new cutie at the office, or maybe they just miss old friends. Most of the names in AOL's logs are too ambiguous to pinpoint to a single person in the real world, so don't get too tweaked if you find your own name and hometown in there. I've got it much worse. There are 36 million searches here, but none of them are for me.

THE SHOPPER

The user who hits "treo 700" 37 times in three days. Here, the data didn't confirm my biases. I'd expected to find window shoppers who searched for Porsche Cayman pages every weekend. But AOL's logs reveal that searches for "coupons" are a lot more common. My favorite specimen is the guy who mostly looked up food brands like Dole, Wendy's, Red Lobster, and Turkey Hill, with an occasional break for "asian movie stars." How much more American could America Online get?

THE OBSESSIVE

The guy who searches for the same thing over and over and over. Looking at the search words themselves can obfuscate a more general long-term pattern—A, A, A, A, B, A, A, C,

A, D, A—that suggests a user who can't let go of one topic, whether it's Judaism, real estate, or Macs. Obsessives are most likely to craft advanced search terms like *craven randy fanfic -wes* and *pfeffern**sse*.

THE OMNIVORE

Many users aren't obsessive—they're just online a lot. My taxonomy fails them, because their search terms, while frequent, show little repetition or regularity. Still, I can spot a few subcategories. There are the trivia buffs who searched "imdb" hundreds of times in three months and the nostalgia surfers on the hunt for "pat benatar helter skelter lyrics."

THE NEWBIE

They just figured out how to turn on the computer. User No. 12792510 is one of many who confuses AOL's search box with its browser address window—he keeps seaching for "www.google." Other AOLers type their searches without spaces between the words ("newcaddillacdeville") as if they were 1990s-era AOL keywords.

THE BASKET CASE

In college I had to write a version of the classic ELIZA program, a pretend therapist who only responds to your problems ("I am sad") with more questions ("Why do you say you are sad?"). AOL Search, it seems, serves the same purpose for a lot of users. I stumbled across queries like "i hate my job" and "why am i so ugly." For me, one log entry stands above the rest: "i hurt when i think too much i love roadtrips i hate my weight i fear being alone for the rest of my life." Me too, 3696023. Me too.

A Non-Programmer's Apology

*Everyone says a great programmer should exercise
their talents to the fullest. But what if programming isn't what you want to do?*

In his classic *A Mathematician's Apology,* published 65 years
ago, the great mathematician G. H. Hardy wrote that "A
man who sets out to justify his existence and his activities"
has only one real defense, namely, that "I do what I do
because it is the one and only thing that I can do at all well."
"I am not suggesting," he added,

> that this is a defence which can be made by most peo-
> ple, since most people can do nothing at all well. But it
> is impregnable when it can be made without absurdity.
> . . . If a man has any genuine talent he should be ready
> to make almost any sacrifice in order to cultivate it to
> the full.

Reading such comments one cannot help but apply them to
oneself, and so I did. Let us eschew humility for the sake of
argument and suppose that I am a great programmer. By
Hardy's suggestion, the responsible thing for me to do
would be to cultivate and use my talents in that field, to

spend my life being a great programmer. And that, I have to say, is a prospect I look upon with no small amount of dread.

It was not always quite this way. For quite a while programming was basically my life. And then, somehow, I drifted away. At first it was small steps—discussing programming instead of doing it, then discussing things *for* programmers, and then discussing other topics altogether. By the time I reached the end of my first year in college, when people were asking me to program for them over the summer, I hadn't programmed in so long that I wasn't even sure I really could. I certainly did not think of myself as a particularly good programmer.

Ironic, considering Hardy writes that

> Good work is not done by "humble" men. It is one of the first duties of a professor, for example, in any subject, to exaggerate a little both the importance of his subject and his own importance in it. A man who is always asking "Is what I do worthwhile?" and "Am I the right person to do it?" will always be ineffective himself and a discouragement to others. He must shut his eyes a little and think a little more of his subject and himself than they deserve. This is not too difficult: it is harder not to make his subject and himself ridiculous by shutting his eyes too tightly.

Perhaps, after I had spent so much time not programming, the blinders had worn off. Or perhaps it was the reverse: that I had to convince myself that I was good at what I was doing now and, since that thing was not programming, by extension, that I was not very good at programming.

Whatever the reason, I looked upon the task of actually having to program for three months with uncertainty and trepidation. For days, if I recall correctly, I dithered. Thinking myself incapable of serious programming, I thought to wait until my partner arrived and instead spend my time assisting him. But days passed, and I realized it would be weeks before he would appear, and I finally decided to try to program something in the meantime.

To my shock, it went amazingly well, and I have since become convinced that I'm a pretty good programmer, if lacking in most other areas. But now I find myself faced with this dilemma: it is those other areas I would much prefer to work in.

The summer before college I learned something that struck me as incredibly important and yet known by very few. It seemed clear to me that the only responsible way to live my life would be to do something that would only be done by someone who knew this thing—after all, there were few who did and many who didn't, so it seemed logical to leave most other tasks to the majority.

I concluded that the best thing to do would be to attempt to explain this thing I'd learned to others. Any specific task I could do with the knowledge would be far outweighed by the tasks done by those I'd explained the knowledge to.[1] It was only after I'd decided on this course of action (and perhaps this is the blinders once again) that it struck me that explaining complicated ideas was actually something I'd always loved doing and was really pretty good at.

That aside, now that I've spent the morning reading David Foster Wallace (DFW),[2] it is plain that I am no great writer. And so, reading Hardy, I am left wondering whether my decision is somehow irresponsible.

I am saved, I think, because it appears that Hardy's logic to some extent parallels mine. Why is it important for the man who "can bat unusually well" to become "a professional cricketer"? It is, presumably, because those who can bat unusually well are in short supply and so the few who are gifted with that talent should do us all the favor of making use of it. If those whose "judgment of the markets is quick and sound" become cricketers, while the good batters become stockbrokers, we will end up with mediocre cricketers and mediocre stockbrokers. Better for all of us if the reverse is the case.

But this, of course, is awfully similar to the logic I myself employed. It is important for me to spend my life explaining what I've learned because people who have learned it are in short supply—much shorter supply, in fact (or so it appears), than people who can bat well.

However, there is also an assumption hidden in that statement. It only makes sense to decide what to become based on what you can presently do if you believe that abilities are somehow granted innately and can merely be cultivated, not created in themselves. This is a fairly common view, although rarely consciously articulated (as indeed Hardy takes it for granted), but not one that I subscribe to.

Instead, it seems plausible that talent is made through practice, that those who are good batters are that way after spending enormous quantities of time batting as kids.[3] Mozart, for example, was the son of "one of Europe's leading musical teachers,"[4] and said teacher began music instruction for his son when the child was three years old. While I am plainly no Mozart, several similarities do seem apparent. My father had a computer programming company, and he began showing me how to use the computer as far back as I can remember.

The extreme conclusion from the theory that there is no innate talent is that there is no difference between people and thus, as much as possible, we should get people to do the most important tasks (writing, as opposed to cricket, let's say). But in fact this does not follow.

Learning is like compound interest. A little bit of knowledge makes it easier to pick up more. Knowing what addition is and how to do it, you can then read a wide variety of things that use addition, thus knowing even more and being able to use that knowledge in a similar manner.[5] And so, the growth in knowledge accelerates.[6] This is why children who get started on something at a young age, as Mozart did, grow up to have such an advantage.

And even if (highly implausibly) we were able to control the circumstances in which all children grew up so as to maximize their ability to perform the most important tasks, that still would not be enough, since in addition to aptitude there is also interest.

Imagine the three sons of a famous football player. All three are raised similarly, with athletic activity from their earliest days, and thus have an equal aptitude for playing football. Two of them pick up this task excitedly, while one, despite being good at it, is uninterested[7] and prefers to read books.[8] It would not only be unfair to force him to use his aptitude and play football, it would also be unwise. Someone whose heart isn't in it is unlikely to spend the time necessary to excel.

And this, in short,[9] is the position I find myself in. I don't *want* to be a programmer. When I look at programming books, I am more tempted to mock them than to read them. When I go to programmer conferences, I'd rather skip out and talk politics than programming. And writing code, although it can be enjoyable, is hardly something I want to spend my life doing.

Perhaps, I fear, this decision deprives society of one great programmer in favor of one mediocre writer. And let's not hide behind the cloak of uncertainty; let's say we know that it does. Even so, I would make it. The writing is too important, the programming too unenjoyable.

And for that, I apologize.

NOTES

1. Explaining what that knowledge is, naturally, is a larger project and must wait for another time.

2. You can probably see DFW's influence on this piece, not least of which in these footnotes.

3. Indeed, this apparently parallels the views of the psychologists who have studied the question. Anders Ericsson, a psychology professor who studies "expert performance," told the *New York Times Magazine* (*NYTM*) that "the most general claim" in his work "is that a lot of people believe there are some inherent limits they were born with. But there is surprisingly little hard evidence." The conclusion that follows, the *NYTM* notes, is that "when it comes to choosing a life path, you should do what you love—because if you don't love it, you are unlikely to work hard enough to get very good. Most people naturally don't like to do things they aren't 'good' at. So they often give up, telling themselves they simply don't possess the talent for math or skiing or the violin. But what they really lack is the desire to be good and to undertake the deliberate practice that would make them better."

4. The quote is from Wikipedia, where, indeed, the other facts are drawn from as well, the idea having been suggested by Stephen Jay Gould's essay "Mozart and Modularity" in his book *Eight Little Piggies*.

5. I've always thought that this was the reason kids (or maybe just me) especially disliked history. Every other field—biology, math, art—had at least some connection to the present, and thus kids had some foundational knowledge to build on. But history? We simply weren't there and thus know absolutely nothing of it.

6. It was tempting to write that "the rate of growth" accelerates, but that would mean something rather different.

7. Many people, of course, are uninterested in such things precisely because they aren't very good at them. There's nothing like repeated failures to turn you away from an activity. Perhaps this is another reason to start young—young children might be less stung by failure, as little is expected from them.

8. I apologize for the clichédness of this example.

9. Well, shorter than most DFW.

Matt Gaffney

The Ultimate Crossword Smackdown

Who writes better puzzles, humans or computers?

When people find out that I write crosswords for a living, they often ask, "Can't you just write crosswords using a computer program now?" After I finish crying—some people really know how to hurt a guy—I respond that, yes, computers play a role in crossword design these days. There are three parts to constructing a crossword: coming up with a theme, filling in the grid, and writing the clues. Until artificial intelligence makes some serious leaps, humans will do the heavy lifting when it comes to theme creation and clue writing. But the second part, filling grids with words, is quite computer friendly. It's here that machines have revolutionized the construction of crossword puzzles.

Early efforts in computer-aided crossword design spat out marginal little grids filled with obscure words. But in the late 1980s, Boston computer programmer Eric Albert had an insight while tangling with this problem: A computer could generate high-quality crossword puzzles if each entry in its word database were ranked on, say, a scale from 1 to 10. An excellent puzzle word like JUKEBOX (gotta love all those high-scoring Scrabble letters) might be worth a 9 or a 10, while a hacky obscurity like UNAU (a type of sloth that has appeared in crosswords more times than it's been spotted in real life) would be a 1 or a 2. By ranking the words, the junk would be left out, and just the good stuff would go in.

This is how computer-aided crossword design still works today. The database operator has to place theme entries and black squares logically in the grid; this placement is done intuitively, based on what the human thinks the computer can handle. After the computer fills in the blanks, the human operator will likely do some further tweaking, such as marking off a corner of the grid he or she doesn't like so the computer can take another shot at it.

The best databases belong to Frank Longo and Peter Gordon, who work in the service of the two best daily crosswords in the country, the *New York Times* and *New York Sun*. Longo's database contains about 720,000 words—pretty much every word or combination of words that's ever been used in a high-quality puzzle. Longo and Gordon use these massive word libraries in concert with their intuition to craft wonderful, tricky puzzles that are enjoyed by multitudes of crossword enthusiasts.

The number of humans who can write a better puzzle than the top databases is small and dwindling. New constructors who come into the field are likely to use computer assistance right off the bat. I'm one of the few holdouts—I don't use a computer database to fill my grids. When am I going to be replaced by a computer? Are my days already numbered?

To answer that, I put together the Ultimate Crossword Smackdown: two top human constructors versus Longo's and Gordon's databases. I humbly chose myself, the author of more than 2,500 crosswords in the past 20 years, to represent humanity. I also selected Santa Clara University math professor Byron Walden, one of the best constructors in the country. The panel of judges will be familiar if you've seen the documentary *Wordplay*. Ellen Ripstein, Tyler Hinman, Trip Payne, and Jon Delfin are all former champions of the American Crossword Puzzle Tournament. The fifth judge is esteemed *Los Angeles Times* crossword editor Rich Norris. They will evaluate the puzzles based on the quality of the words used, the amount of "crosswordese"—UNAU-type words you never see anywhere but the puzzle page—and what can best be described as the puzzle's "overall feel." These folks know a good crossword when they see one.

We all start with a standard 15-by-15 square crossword puzzle grid. I've created the puzzle's theme (famous people with the initials B.B.) and placed the three 15-letter theme entries, which are unchangeable and unmovable.

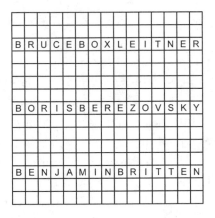

Longo and Gordon feed the theme entries into their computers and then place black squares into the grid themselves based on their sense of what pattern their databases can handle. Walden and I likewise scatter a few black squares around the grid, and we're off to the races.

My strategy is to fit as many *X*'s, *Q*'s, *J*'s, and *Z*'s in as possible. Hopefully, I won't get penalized for the crosswordese that creeps in to accommodate those high-value letters.

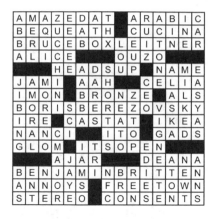

Overall, I'm fairly happy with my grid. The puzzle has some lousy entries like ATO (as in "From ___ Z," a legit-but-dull crossword standard), EEEEE ("Widest male shoe size"), and INE ("Chemistry suffix"). There are some really nice entries, though, like THX (short for "thanks") and ZZ TOP and ZUCCHINI and SHORT *I*'S ("Some vowels").

And here are the other three entries. For now, I'll keep you in the dark about which one emerged from the mind of Byron Walden and which ones came from the computers. If you'd like to play along, give four points to the puzzle you think is best, three to your second favorite, and so forth.

GRID A:

A	B	B	E	■	L	E	P	E	W	■	A	B	I	T
M	A	R	X	■	I	R	A	N	I	■	R	E	N	O
B	R	U	C	E	B	O	X	L	E	I	T	N	E	R
I	N	T	U	X	E	D	O	S	■	D	I	Z	Z	Y
■	■	S	I	R	E	S	■	L	I	C	■			
A	L	W	E	S	T	■	B	I	G	L	I	A	R	
L	E	I	■	T	Y	P	E	I	N	■	E	T	N	A
B	O	R	I	S	B	E	R	E	Z	O	V	S	K	Y
O	X	E	N	■	O	P	E	N	E	R	■	U	L	E
M	I	D	J	U	N	E	■	R	S	V	P	E	D	
■	■	U	N	D	■	S	A	T	I	E	■			
B	R	O	N	C	■	L	E	M	O	N	S	O	L	E
B	E	N	J	A	M	I	N	B	R	I	T	T	E	N
L	E	T	O	■	O	N	S	E	T	■	R	O	A	D
S	L	O	E	■	D	E	E	R	E	■	Y	E	N	S

GRID B:

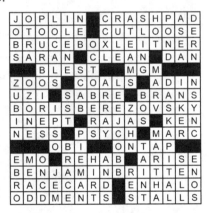

D	A	N	M	A	R	I	N	O	█	D	A	N	T	E
A	R	E	A	C	O	D	E	S	█	A	L	O	H	A
B	R	U	C	E	B	O	X	L	E	I	T	N	E	R
B	I	T	E	█	█	T	O	G	S	█	U	S	C	
E	V	E	R	A	G	E	█	█	G	E	I	S	H	A
D	E	R	█	L	A	U	D	E	█	S	H	E	E	N
█	█	W	A	R	G	A	M	E	█	A	R	I	D	
B	O	R	I	S	B	E	R	E	Z	O	V	S	K	Y
A	V	I	D	█	O	N	E	T	I	M	E	█	█	
D	E	D	E	E	█	E	D	I	N	A	█	M	L	B
D	R	E	N	C	H	█	█	C	E	N	T	A	U	R
E	L	I	█	L	E	G	G	█	█	I	N	C	A	
B	E	N	J	A	M	I	N	B	R	I	T	T	E	N
T	A	T	E	S	█	R	A	T	I	S	L	A	N	D
S	P	O	T	S	█	D	R	U	G	T	E	S	T	S

GRID C:

J	O	P	L	I	N	█	C	R	A	S	H	P	A	D
O	T	O	O	L	E	█	C	U	T	L	O	O	S	E
B	R	U	C	E	B	O	X	L	E	I	T	N	E	R
S	A	R	A	N	█	C	L	E	A	N	█	D	A	N
█	█	B	L	E	S	T	█	█	M	G	M	█	█	
Z	O	O	S	█	C	O	A	L	S	█	A	D	I	N
U	Z	I	█	S	A	B	R	E	█	B	R	A	N	S
B	O	R	I	S	B	E	R	E	Z	O	V	S	K	Y
I	N	E	P	T	█	R	A	J	A	S	█	K	E	N
N	E	S	S	█	P	S	Y	C	H	█	M	A	R	C
█	█	O	B	I	█	O	N	T	A	P	█	█		
E	M	O	█	R	E	H	A	B	█	A	R	I	S	E
B	E	N	J	A	M	I	N	B	R	I	T	T	E	N
R	A	C	E	C	A	R	D	█	E	N	H	A	L	O
O	D	D	M	E	N	T	S	█	S	T	A	L	L	S

The judges' opinions were pretty scattershot. Each of the four participants got at least one last-place vote, and three of the four contestants got at least one first-place vote. Most amazingly, two of the judges ranked the four grids in the precise reverse order of each other. There's clearly a sizable dose of subjectivity in assessing the quality of a crossword puzzle, especially at this, ahem, elite level.

The final results: GRID A prevailed with 14 points. Close behind were GRID B and GRID C, each with 13 points. I brought up the caboose (sigh) with 10 points.

In GRID A, the judges liked INJUN JOE, LEMON

SOLE, I DIG, and ARTICLE V. IN TUXEDOS, though, was slammed as a contrived phrase. That didn't keep the puzzle from taking two of the five first-place votes.

In GRID B, judges liked DAN MARINO, EAR CANDY, and EZINE. They weren't partial to RAT ISLAND, which was deemed a little obscure.

GRID C got props for LEE J. COBB, N SYNC, and CRASH PAD. But POURBOIRES, a French word meaning "tips," as in a restaurant or bar, was cited by two judges as being outside their ken.

My grid drew the praise and condemnation I expected: one first-place vote but two last places as well. One judge wrote: "This puzzle, to me, screams 'trying too hard.'"

And now for the big reveal. GRID A (14 points) was written by Byron Walden, GRID B (13 points) by Frank Longo's database, and GRID C (13 points) by Peter Gordon's database. That makes the final score: Computers 26, Humans 24. I tip my hat to our cyberchampions but will not hesitate to unplug them if they get too smug in victory. They can't get too smug anyway, because Byron's top score is a reassuring note for human constructors.

No sample test proves anything with certainty. Different competitors, different judges, a different day—anything could've pushed the results one way or the other. Still, the computers won this battle on points, and they have other advantages that make the future ominous for human constructors. Their databases are getting larger every year, and they work much more quickly than we do. A typical human constructor may take between two and five hours to fill a challenging 15-by-15 grid. Longo's and Gordon's databases can sometimes do it in under a minute.

Will I be jumping onboard the database ship? It may be a poor career move, but I will not. I'd rather figure out how to win a rematch than join the other team.

Lori Gottlieb

How Do I Love Thee?

A growing number of Internet dating sites are relying on academic researchers to develop a new science of attraction. A firsthand report from the front lines of an unprecedented social experiment.

I'd been sitting in Dr. Neil Clark Warren's office for less than 15 minutes when he told me he had a guy for me. It wasn't surprising that the avuncular 71-year-old founder of eHarmony.com, one of the nation's most popular online dating services, had matchmaking on his mind. The odd thing was that he was eager to hook me up without having seen my eHarmony personality profile.

I'd come to the eHarmony headquarters in Pasadena, California, in early October to learn more about the site's "scientifically proven" and patented Compatibility Matching System. Apparently, the science wasn't working for me. The day before, after I'd taken the company's exhaustive (and exhausting) 436-question personality survey, the computer informed me that of the approximately 9 million eHarmony members, more than 40 percent of whom are men, I had zero matches. Not just in my city, state, region, or country but in the entire world. So Warren, who looks like Orville Redenbacher and speaks with the folksy cadence of Garrison Keillor, suggested setting me up with one of his company's advisory board members, whom he described as brilliant, Jewish, and 38 years old. According to Warren, this board member, like me, might have trouble finding a match on eHarmony.

"Let me tell you why you're such a difficult match," Warren said, facing me on one of his bright floral sofas. He started running down the backbone of eHarmony's predictive model of broad-based compatibility, the so-called 29 dimensions (things like curiosity, humor, passion, intellect), and explaining why I and my prospective match were such outliers.

"I could take the 9 million people on our site and show you dimension by dimension how we'd lose people for you," he began. "Just on IQ alone—people with an IQ lower than 120, say. Okay, we've eliminated people who are not intellectually adequate. We could do the same for people who aren't creative enough or don't have your brilliant sense of humor. See, when you get on the tails of these dimensions, it's really hard to match you. You're too bright. You're too thoughtful. The biggest thing you've got to do when you're gifted like you are is to be patient."

After the over-the top flattery wore off—and I'll admit, it took an embarrassingly long time—I told Warren that most people I know don't join online dating sites to be patient. Impatience with real-world dating, in fact, is precisely what drives many singles to the fast-paced digital meat market. From the moment Match.com, the first such site, appeared in 1995, single people suddenly had 24-hour access to thousands of other singles who met their criteria in terms of race, religion, height, weight, even eye color and drinking habits.

Nearly overnight, it seemed, dozens of similar sites emerged, and online dating became almost de rigueur for busy singles looking for love. According to a recent Pew survey, 31 percent of all American adults (63 million people) know someone who has used a dating Web site, while 26 percent (53 million people) know someone who has gone out with a person he or she met through a dating Web site. But was checking off boxes in columns of desired traits, like an à la carte Chinese take-out menu, the best way to find a soul mate?

Enter eHarmony and the new generation of dating sites, among them PerfectMatch.com and Chemistry.com. All have staked their success on the idea that long-term romantic compatibility can be predicted according to scientific principles—and that they can discover those principles and use them to help their members find lasting love. To that end they've hired high-powered academics, devised special algorithms for relationship-matching, developed sophisticated personality questionnaires, and put into place mechanisms for the long-term tracking of data. Collectively, their efforts mark the early days of a social experiment of unprecedented proportions, involving millions of couples and possibly extending over the course of generations. The question at the heart of this grand trial is simple: In the subjective realm of love, can cold, hard science help?

Although eHarmony was the first dating site to offer science-based matching, Neil Clark Warren seems like an unlikely pioneer in the field. Even though he earned a Ph.D. in clinical psychology from the University of Chicago, in 1967, he never had much of a passion for academic research—or an interest in couples. "I was scared to death of adults," he told me. "So I did child therapy for a while." With a master's degree in divinity from Princeton Theological Seminary, he went on to Fuller Theological Seminary's

Graduate School of Psychology, in southern California, where he taught and practiced humanistic psychology (what he calls "client-centered stuff") in the vein of his University of Chicago mentor, Carl Rogers. "I hated doing research," he admitted, before adding with a smile, "In fact, I was called 'Dr. Warm.'"

Fittingly, it was Warren's family, not academia, that piqued his interest in romantic compatibility. "When my daughters came along, that was a big pivot in my life in thinking about how do two people get together," he told me. "I started reading in the literature and realizing what a big chance they had of not having a satisfying marriage. I started trying to look into it."

Soon he began a private practice of couples therapy—with a twist. "People have always thought, wrongly, that psychotherapy is a place to go deal with problems," he said. "So when a couple would come in, I'd say, 'Tell me how you fell in love. Tell me the funniest thing that's happened in your marriage.' If you want to make a relationship work, don't talk about what you find missing in it! Talk about what you really like about it."

Warren is a big proponent of what he likes to call "folksy wisdom." One look at the shelves in his office confirms this. "I've been reading this little book about the Muppets—you know, Jim Henson," he said. "And I've been reading another book about Mister Rogers. I mean, Mister Rogers was brilliant beyond belief! He got a hold of concepts so thoroughly that he could transmit them to 6-year-old kids! Do you know how much you have to get a hold of a concept to transmit it simply? His idea of simple-but-profound has had a profound influence on me."

The basis of eHarmony's matching system also sounds simple but profound. In successful relationships, Warren says, "similarities are like money in the bank. Differences are like debts you owe. It's all right to have a few differences, as long as you have plenty of equity in your account."

He leaned in and lowered his voice to a whisper. "Mister Rogers and Jim Henson," Warren continued, "they got a hold of the deep things of life and were able to put them out there. So that's what we want to do with our products. We want to put them out there in a way that you'd say, 'This is common sense. This seems right, this seems like it would work.' Our idea of broad-based compatibility, I put it out there in front of you. Does that seem right?"

Whether or not it seems right on an intuitive level is almost beside the point. After all, eHarmony's selling point, its very brand identity, is its scientific compatibility system. That's where Galen Buckwalter comes in.

A vice president of research and development for the company, Buckwalter is in charge of recruiting what he hopes will be 20 to 25 top relationship researchers away from academia—just as he was lured away by Warren 9 years ago. A former psychology graduate student at Fuller Theological Seminary (his dissertation was titled "Neuropsychological Factors Affecting Survival in Malignant Glioma Patients Treated with Autologous Stimulated Lymphocytes"), Buckwalter had become an assistant professor at the University of Southern California, where he was studying the effects of hormones on cognition, when he got the call from Warren.

"Neil knew I lived and breathed research, and he had this idea to try to develop some empirically based model to match people," Buckwalter said when I visited him at his office at eHarmony. He wore a black T-shirt and wire-rimmed glasses and had a hairstyle reminiscent of Einstein's. "He wasn't necessarily thinking over the Internet— maybe a storefront operation like Great Expectations."

Relationships weren't Buckwalter's area, but he welcomed the challenge. "A problem is a problem, and relationships are a good problem," he said. "In the research context, it's certainly an endlessly fascinating question."

With the help of a graduate student, Buckwalter reviewed the psychological literature to identify the areas that might be relevant in predicting success in long-term relationships. "Once we identified all those areas, then we put together a questionnaire—just a massively long questionnaire," he said. "It was probably close to a thousand questions. Because if you don't ask it, you're never gonna know. So we had tons of questions on ability, even more on interest. Just every type of personality aspect that was ever measured, we were measuring it all."

Because it wasn't practical to execute a 30-year longitudinal study, he and Warren decided to measure existing relationships, surveying people who were already married. The idea was to look for patterns that produce satisfaction in marriages and then try to reproduce them in the matching of singles.

Buckwalter's studies soon yielded data that confirmed one of Warren's longtime observations: namely, that the members of a happy couple are far more similar to each other than are the members of an unhappy couple. Compatibility, in other words, rests on shared traits. "I can't tell you how delighted I was," Warren said, "when the factor-analytic studies started bringing back the same stuff I'd seen for years."

But could this be true across the board? I told Warren that my most successful relationships have been with men who are far less obsessive than I am. Warren assured me that's not a similarity their system matches for. "You don't want two obsessives," he explained. "They'll drive each other crazy. You don't find two control freaks in a great marriage. So we try to tweak the model for that. Fifty percent of the ball game is finding two people who are stable."

For Warren, a big question remained: What should be done with these findings? Originally, he had partnered with his son-in-law, Greg Forgatch, a former real-estate developer, to launch the business. Their first thought was to produce educational videotapes on relationship compatibility. After all, Warren had recently written his book, *Finding the Love of Your Life*.

"We tried so hard to make videotapes and audiotapes," Warren said. "I went into the studio and made lists. We came up with a hundred things singles need. But singles don't want education; they want flesh! They want a person. So that's when, in 1997, we said, 'We've gotta help people find somebody who would be good for them. Some *body*.'"

To connect singles and create a data pool for more research, the Internet seemed the best option. Based on a study of 5,000 married couples, Warren put together the compatibility model that became the basis for eHarmony. "We got encouraged by everybody, 'Get out there, get out there! The first person to market is going to be the most successful,'" Warren recalled. But he insisted on getting the matching system right before launching the site—and that didn't happen until August 2000, during the dot-com bust. By 2001 he was contemplating declaring bankruptcy.

"And then," Warren recalled, "we found an error in our matching formula, so a whole segment of our people were not getting matched. It was an error with all the Christian people on the site."

This is a sensitive topic for Warren, who bristles at the widely held opinion that eHarmony is a Christian dating site. The company's chief operating officer, he offered by way of rebuttal, is Jewish, and Buckwalter, who became a quadriplegic at age 16 after jumping into a river and break-

ing his neck, is agnostic. And while Warren describes himself as "a passionate Christian" and proudly declares, "I love
Jesus," he worried about narrowing the site with too many
questions about spiritual beliefs. Which is where the error
came in.

"We had seven questions on religion," he explained,
"and we eliminated four of them. But we forgot to enter that
into the matching formula! These were 7-point questions.
You needed 28 points to get matched with a Christian person, but there was no way you could get them! We only had
three questions! So every Christian person who had come to
us had zero matches."

Fortunately, a wave of positive publicity, featuring married couples who'd met through eHarmony and the naturally charismatic Warren, turned things around. Still, Warren said of the innocent mistake, "you kind of wonder how
many relationships fall apart for reasons like this—how
many businesses?"

Today, eHarmony's business isn't just about using science to
match singles online. Calling itself a "relationship-enhance-
ment service," the company has recently created a venture-
capital-funded think tank for relationship and marital
research, headed up by Dr. Gian Gonzaga, a scientist from
the well-known marriage-and-family lab at the University
of California at Los Angeles. The effort, as Gonzaga put it
to me recently, is "sort of like a Bell Labs or Microsoft for
love."

An energetic, attractive 35-year-old, Gonzaga thought
twice about leaving the prestige of academia. "It seemed
cheesy at first," he said. "I mean, this was a dating service."
But after interviewing with Warren, he realized that conducting his research under the auspices of eHarmony would
offer certain advantages. He'd be unfettered by teaching and

grant writing, and there would be no sitting on committees or worrying about tenure. More important, since his research would now be funded by business, he'd have the luxury of doing studies with large groups of ready subjects over many years—but without the constraints of having to produce a specific product.

"We're using science in an area most people think of as inherently unscientific," Gonzaga said. So far, the data are promising: a recent Harris Interactive poll found that between September 2004 and September 2005, eHarmony facilitated the marriages of more than 33,000 members—an average of 46 marriages a day. And a 2004 in-house study of nearly 300 married couples showed that people who met through eHarmony report more marital satisfaction than those who met by other means. The company is now replicating that study in a larger sample.

"We have massive amounts of data!" Warren said. "Twelve thousand new people a day taking a 436-item questionnaire! Ultimately, our dream is to have the biggest group of relationship psychologists in the country. It's so easy to get people excited about coming here. We've got more data than they could collect in a thousand years."

But how useful is this sort of data for single people like me? Despite Warren's disclaimer about what a tough eHarmony match I am, I did finally get some profiles in my inbox. They included a bald man with a handlebar moustache, who was 14 inches taller than me; a 5-foot-4-inch attorney with no photos; and a film editor whose photo shows him wearing a kilt—and not in an ironic way. Was this the best science could do?

When I asked Galen Buckwalter about this, he laughed, indicating that he'd heard the question before. "The thing you have to remember about our system is we're matching on these algorithms for long-term compatibility," he said.

"Long-term satisfaction is not the same as short-term attraction. A lot of people, when they see their initial matches, it's like, 'This is crap!' "

In ads and on his Web site, Warren talks about matching people "from the inside out." Was eHarmony suggesting that I overlook something as basic as romantic chemistry? "When we started out," Buckwalter said, "we were almost that naive." But now, he added, eHarmony is conducting research on the nature of physical attraction.

"We're trying to find out if we can predict physical chemistry with the same degree of statistical certainty that we've used to predict long-term satisfaction through our compatibility matching. In general, people seem to be attracted to people who share their physical attributes," Buckwalter explained, noting that he has found some exceptions, like height preference. "There's a lot of variability on that dimension," he said. "A person's height, it turns out, is not a consistent predictor of short-term attraction." Meanwhile, Buckwalter's team is in the process of testing new hypotheses.

"We're still convinced that our compatibility-matching process is essential for long-term satisfaction, so we're not going to mess with that," he insisted. "But if we can fit a short-term attraction model on top of that, and it's also empirically driven, that's the Holy Grail."

Over at Chemistry.com, a new site launched by Match.com, short-term attraction is already built into the system. This competitor of eHarmony's was developed with help from Match.com's chief scientific adviser, Dr. Helen Fisher, an anthropologist at Rutgers University, whose research focuses on the brain physiology of romantic love and sexuality. Chemistry.com is currently assembling a multidisciplinary group of psychologists, relationship counselors, sociologists, neuroscientists, and sexologists to serve as consultants.

The company sought out Fisher precisely because its market research revealed that although a large segment of singles wanted a scientific approach, they didn't want it to come at the expense of romantic chemistry. "On most of the other sites, there's this notion of 'fitness matching,'" Fisher said from her office in New York City. "You may have the same goals, intelligence, good looks, political beliefs. But you can walk into a room, and every one of those boys might come from the same background, have the same level of intelligence, and so on, and maybe you'll talk to three but won't fall in love with any of them. And with the fourth one, you do. What creates that chemistry?"

It's a constellation of factors, Fisher told me. Sex drive, for instance, is associated with the hormone testosterone in both men and women. Romantic love is associated with elevated activity of the neurotransmitter dopamine and probably also another one, norepinephrine. And attachment is associated with the hormones oxytocin and vasopressin. "It turns out," she said, "that seminal fluid has all of these chemicals in it. So I tell my students, 'Don't have sex if you don't want to fall in love.'"

Romantic love, Fisher maintains, is a basic mating drive—more powerful than the sex drive. "If you ask someone to go to bed with you, and they reject you," she says, "you don't kill yourself. But if you're rejected in love, you *might* kill yourself."

For Chemistry.com's matching system, Fisher translated her work with neurotransmitters and hormones into discrete personality types. "I've always been extremely impressed with Myers-Briggs," she said, referring to the personality assessment tool that classifies people according to four pairs of traits: Introversion versus Extroversion,

Sensing versus Intuition, Thinking versus Feeling, and Judging versus Perceiving. "They had me pinned to the wall when I took the test, and my sister, too. So when Chemistry.com approached me, I said to myself, 'I'm an anthropologist who studies brain chemistry; what do I know about personality?'"

Turns out she knew quite a bit: Genes for the activity of dopamine are associated with motivation, curiosity, anxiety, and optimism. Genes for the metabolism of serotonin, another neurotransmitter, tend to modulate one's degree of calm, stability, popularity, and religiosity. Testosterone is associated with being rational, analytical, exacting, independent, logical, rank oriented, competitive, irreverent, and narcissistic. And the hormone estrogen is associated with being imaginative, creative, insightful, humane, sympathetic, agreeable, flexible, and verbal.

"So I had these four sheets of paper," Fisher continued. "And I decided to give each a name. Serotonin became the Builder. Dopamine, the Explorer. Testosterone, the Director. And estrogen—I wish I'd called it the Ambassador or Diplomat, but I called it the Negotiator." Myers-Briggs, she says, "clearly knew the four types but didn't know the chemicals behind them."

The 146-item compatibility questionnaire on Chemistry.com correlates users' responses with evidence of their levels of these various chemicals. One question, for instance, offers drawings of a hand, then asks:

Which one of the following images most closely resembles your left hand?
Index finger slightly longer than ring finger
Index finger about the same length as ring finger
Index finger slightly shorter than ring finger
Index finger significantly shorter than ring finger

The relevance of this question might baffle the average online dater accustomed to responding to platitudes like "How would you describe your perfect first date?" But Fisher explains that elevated fetal testosterone determines the ratio of the second and fourth finger in a particular way as it simultaneously builds the male and female brain. So you can actually look at someone's hand and get a fair idea of the extent to which they are likely to be a Director type (ring finger longer than the index finger) or a Negotiator type (index finger longer or the same size).

Another question goes like this:

How often do you vividly imagine extreme life situations, such as being stranded on a desert island or winning the lottery?
Almost never
Sometimes
Most of the time
All the time

"Someone who answers 'All the time' is a definite Negotiator," Fisher said. "High estrogen activity is associated with extreme imagination."

While other sites gather data based on often unreliable self-reports ("How romantic do you consider yourself to be?"), many of the Chemistry.com questions are designed to translate visual interpretation into personality assessment, thus eliminating some of the unreliability. In one, the user is presented with a book's jacket art. We see a woman in a sexy spaghetti-strapped dress gazing at a man several feet away in the background, where he leans on a stone railing. The sky is blue, and they're overlooking an open vista. "What is the best title for this book?" the questionnaire asks, and the choices are as follows:

A Spy in Rimini
Anatomy of Friendship: A Smart Guide for Smart People
A Scoundrel's Story
Things Left Unsaid

According to Fisher, each response is correlated with one of the four personality types: Choice A corresponds to Explorer, B to Builder, C to Director, and D to Negotiator.

Even sense of humor can be broken down by type, with questions like "Do you sometimes make faces at yourself in the mirror?" (people with a sense of humor do) and "At the zoo, which do you generally prefer to watch?" (the reply "monkeys and apes" indicates more of a funny bone than "lions and tigers"). According to Fisher, a Director likes people to laugh at his or her jokes; a Negotiator likes to be around someone funny so he or she can laugh at that person's jokes; an Explorer is spontaneous and laughs at just about anything; and a Builder, she suspects, generally isn't as funny as the others.

But how to match people up according to Fisher's four personality types, and under what circumstances, isn't so straightforward. Another question, for instance, presents four smiling faces and asks:

Take a look at the faces below. Are their smiles sincere?

Fisher says that people with high levels of estrogen—usually women—have better social skills, and are better at reading other people. So users who choose the correct "real" smiles (pictures two and three) will be the Negotiators. This, Fisher says, is an area where "complementarity" might be important. The problem with sites like eHarmony, she believes, is that they place too much emphasis on similarity, whereas, in her view, falling in love depends on two elements: similarity and complementarity. "We also want someone who masks our flaws," she explained. "For example, people with poor social skills sometimes gravitate toward people with good social skills. I'm an Explorer, so I don't really need a partner who is socially skilled. That's not essential to me. But it may be essential to a Director, who's generally less socially skilled."

Chemistry.com's compatibility questionnaire also examines secondary personality traits. To illustrate, Fisher cited her own relationship. "I'm currently going out with a man," she said, "and of course I made him take the test instantly. We're both Explorers and older. I'm not sure two Explorers want to raise a baby together, because nobody will be home. But in addition, I'm a Negotiator and he's a Director type. Our dominant personality is similar, but underneath, we're complementary."

Determining which works best—similarity or complementarity—may change with the circumstances. A young woman who's an Explorer, Fisher said, might be attracted to a Builder, someone who's more of a homebody, loyal, dependable, and protective. But the pair will be more compatible if their secondary personalities match—maybe they're both Negotiators underneath.

"Nobody is directly locked into any one of these temperament types," Fisher said. "That's why we provide each person with both a major and a minor personality profile. Do Explorers go well together? Do likes attract likes? Sometimes they do, and sometimes they don't."

If this sounds a bit, well, unscientific, Fisher is the first to admit it. "I have theories about what personality type a person would be most ideally suited with," she told me, "but I also trust people to tell me what they are looking for. All throughout the questionnaire are checks and balances to what are just Helen Fisher's theories."

This is why she decided to include an item on the Chemistry.com questionnaire that asks about the traits of a person's partner in his or her most successful former relationship: Was that person an Explorer, a Builder, a Director, a Negotiator? "Anybody can match somebody for values. But I'm hoping to create a system so that five years later they still fascinate each other."

At the same time, Fisher wants couples to be fascinated by each other early on. In other words, why waste time e-mailing back and forth to get to know a potential match over the course of several weeks, as eHarmony encourages its users to do, if there won't be any chemistry when they finally meet? Chemistry.com's guided 1–2–3-Meet system provides a step-by-step structure to get couples face-to-face as soon as possible for that all-important "vibe check." Then there's a postmeeting "chemistry check," where each person offers feedback about the date.

The goal is to incorporate this information into the algorithm to provide better matches, but it can also serve as an accuracy check of the data. Say, for instance, that Jack describes himself as a fashionable dresser, but Jill reports that he showed up for their date in flip-flops, cutoffs, and a do-rag. If the feedback from a number of Jack's first meetings indicates the same problem, Chemistry.com will send him an e-mail saying, "Jack, wear a pair of trousers."

When I asked Helen Fisher how the site's scientific algorithm might change based on this user feedback, she said that perhaps the computer could pick up cues about a person's physical type based on the people he or she finds attractive or unattractive and then send that person closer matches. Or, it might know better than to match me—an avid reader attracted to literary types—with the guy whose personality assessment indicates a literary bent but whose essay reads as follows.

> While I do read books, I have a notoriously short attention span for them. As a result, partially read copies of numerous really good (so I'm told) books are scattered around my apartment. When these get set aside, it's because I've gotten sucked into magazines . . . Every few days, the magazines lose out to DVDs.

It's also possible that user feedback could change the matching formula completely. "We always look at data," Fisher said. "If we find that Explorer/Builder to Director/Negotiator is working for more people, if we find the biochemistry is stronger, we'll adjust that in the formula." Fisher acknowledged that the system right now is mostly a learning tool—a way to collect large amounts of data, look for patterns, and draw conclusions based on the findings.

Still, even a thoroughly researched biochemical model won't prevent glitches in the matching system. In Fisher's view, for example, no scientifically based site would pair her with the men she's dated, because, as she put it, "they're all better looking than me."

"It would be preposterous for anyone to say they can create a formula that works perfectly," she said emphatically. "But I do believe that science can help us get close and that there's a lot more to be learned."

"This test doesn't pretend to be about chemistry," said Dr. Pepper Schwartz—who developed the Duet Total Compatibility System in conjunction with the two-year-old site

PerfectMatch.com. She was speaking by cell phone from San Francisco, where she had just attended a meeting of the National Human Sexuality Resource Center, on whose board she sits. "The chemistry test at Match—that's not about chemistry either. If I could concoct a test for chemistry, I'd make a zillion dollars."

A sociologist at the University of Washington in Seattle, Schwartz is PerfectMatch.com's hipper version of Neil Clark Warren: the accessible, empathic, media-savvy love doctor who guides users through the treacherous dating trenches and onto the path of true compatibility.

According to the site—which calls her by the cutesy moniker "Dr. Pepper"—Schwartz is "the leading relationship expert in the nation," a woman who "holds the distinction of being the only relationship expert on the Web who's a published authority, as well as a professor at a major U.S. university." Oh, and then there are her appearances on *Oprah, The Today Show,* and *Good Morning America;* the 14 books she's written; and her regular column for LifetimeTV.com.

Unlike Warren, however, she neither founded the company (she was brought in by PerfectMatch's Duane Dahl) nor follows Warren's credo of simplicity. In fact, the nifty-sounding Duet Compatibility Profiler takes some complex deconstruction. This makes sense, given that Schwartz has been studying gender relations since the early 1970s, when she was a sociology graduate student at Yale and wrote a PhD thesis on how people hooked up in the college mixer system.

Like Helen Fisher, the Rutgers anthropologist, Schwartz believes that both similarity and complementarity are integral to romantic compatibility. But while Fisher has more of an "it depends" attitude on the question of which of the two makes sense for a particular couple under particular circumstances, Schwartz has a more elaborately defined system, which she outlines in her latest book, *Finding Your Perfect Match*.

Schwartz's Duet model consists of a mere 48 questions and focuses on eight specific personality characteristics: romantic impulsivity, personal energy, outlook, predictability, flexibility, decision-making style, emotionality, and self-nurturing style. On the first four, she believes, a well-suited couple should be similar; on the last four, however, a couple can thrive on either similarity or difference—provided that both people know themselves well enough to determine which works best.

"My first thought was, *Know yourself,*" Schwartz said of how she created PerfectMatch's system. "How can you pick somebody else if you have no insight into yourself?"

Her questionnaire, she believes, will help users to think in a conscious way about who they are. As an example of the kind of introspection she hopes for, Schwartz cites the area of money. "It's a very important thing," she said, "and there's very little research on it, because nobody wants to talk about money. I can ask people if they're orgasmic, and they'll tell me in a second. But ask a subject about money, and they're embarrassed."

When it comes to money, PerfectMatch asks users to get specific—and honest—about how important it is to them. "I want them to think about things like, Should parents pay for college education no matter what it costs? Do you feel you need to make extravagant purchases every once in a while?" Other tests generally stop at innocuous questions about whether people consider themselves fiscally responsible, but Schwartz ventures into un-PC territory with true-or-false statements like "All other things being equal, I tend to respect people who make a lot of money more than people who have modest incomes"; "I could not love a person who doesn't make enough money to help me live the lifestyle I need in order to be happy"; and "I would very much prefer to be with someone who did not have major economic responsibilities to children or parents unless they had a lot of money and these responsibilities did not affect our life together."

Like Chemistry.com's system, Duet has its roots in the Myers-Briggs Type Indicator. But, Schwartz explained, Duet is different from Myers-Briggs in several ways. It has eight characteristics to Myers-Briggs's four; it uses two personality profiles—similarity and complementarity—instead of one; and it relies on studies from a number of fields, rather than just psychology, to determine how these personality characteristics combine in romantic situations, as opposed to general workplace or team-building ones.

"If, say, I'm rigid in my tastes but I have a sense of humor," Schwartz explained, "you can work with me. But if I'm rigid and very earnest, it's going to be difficult. So in our test it's not just, 'Is this person rigid?' Because rigidity can coexist with humor or earnestness, and which one of those traits is present makes a big difference. It's important how these traits are put together."

I took the Duet test and was classified on the similarity scale as X, A, C, and V—that is, Risk Averse, High Energy, Cautious, and Seeks Variety. The site then interpreted the findings, which, to my surprise, rather accurately captured my personality:

> You are careful about entering a relationship. You have a cautious side to your personality on more than one dimension, and so it takes you awhile to believe in love and romance with someone you are dating. Nonetheless, you are a high energy, intense kind of person. Once you believe in a relationship, you can be a good

partner, IF you give it enough time. You demand a lot from the world and you take on a lot. You probably want someone who does the same, or at least supports your own high energy, explorative approach to life.

Yet the complementarity section of my test results—those traits on which my best match might be similar or different—reflected my temperament on only two of the four parameters. I was characterized as S, C, T, and E—that is, Structured, Compromiser, Temperate, and Extrovert—but I'm neither a C nor a T.

Schwartz wasn't ruffled by these inaccuracies. "Perfect-Match is the only scientific site out there that's completely transparent and user operated," she said. "If you disagree with me, you can retake the test anytime and get a different profile that more accurately reflects the subtleties we may have missed. Or you can keep the same profile, but in addition to the matches we provide for you, you can do a search on your own. Say I think a passionate person would want another passionate person. But maybe you know about yourself that you're passionate but want a calm person, someone who stops the escalation of things. I don't care if what you think is theoretically sound; if it doesn't work for you, you can search using your own criteria."

This, she said, distinguishes PerfectMatch from eHarmony and Chemistry.com. "In the Chemistry test," said Schwartz, who is a friend of Helen Fisher's and a fan of her work, "there was a question about where you'd like to live. And I chose the country. And I would—but the people I tend to prefer are in the city. So they sent me people from Bass Breath, Arizona. And there was no way I could change it! At PerfectMatch, we don't overdetermine people's answers that way."

What Schwartz is referring to, of course, is the bugaboo

of all these compatibility-matching systems: nuance. "Even if a site lets you choose physical characteristics like height," she said, "there's no way it's going to guess your physical template. It could be lankiness in one case; it could be somebody's eyes in another. We can't get that out of a questionnaire. Nobody can. So we say, 'Go look at the pictures on our site, see who you find attractive, then look at their personality types and see if they're compatible [with you]. You have that option on PerfectMatch.'"

The advantage to scientific matching, she says, isn't to come up with some foolproof formula for romantic connection. Instead, the science serves as a reality check, as a way of not letting that initial rush of attraction cloud your judgment when it comes to compatibility.

"I went out with a man for about a year who, if I'd taken the test with him—we both would have known we should have stopped early on," Schwartz said. "But, of course, I was attracted to him, and probably to the characteristics that were wrong for me, for the wrong reasons. That's what attraction can do. But if you're also armed with information about compatibility—or lack of compatibility—from the very beginning, you might think twice before getting involved, before you make the mistake of e-mailing the cute guy in the picture, like you might on Match."

Schwartz, who had been married for 23 years before she reentered the dating pool, empathizes with PerfectMatch users. "I know what dating is like," she said. "I'm doing it, too. You start to burn out, and you need to find a certain amount of positive reinforcement. So if we can cut down the really inappropriate personalities for you, we can help out."

Of course, before the days of Myers-Briggs and Perfect-Match and academic departments devoted to deconstructing romantic relationships, there were matchmakers. And today, despite the science, they're still thriving. One of the West Coast's largest matchmaking agencies is called Debra Winkler Personal Search, and its slogan is the opposite of scientific: "The art of the perfect match." Indeed, in the FAQ section of the company's Web site, the reply to "How do you go about matching members?" reads as follows.

Our matchmakers use a combination of tools—including experience and intuition—when matching members. We start with basic demographic information such as age, religion, location, physical requirements and other preferences. Personality profiles are also used but not relied upon exclusively. In the end, however, it comes down to your personal matchmaker.

They hand-select the individuals for you to meet. And it is not based on some absolute, statistical formula. It's more like a feeling, gained from years of experience, that tells them you and another person would be great for each other.

Winkler founded the company 18 years ago and sold it in 2003, leaving its day-to-day operations to Annie Ahlin, who worked with Winkler for 14 years and until November was the company's president.

"Intuition is a big part of determining long-term compatibility," Ahlin told me. She said many of the agency's clients are people who have tried scientific matching online but had no luck. Ahlin believes she knows why. "When you're reading a profile online, or looking at a photo, it's one dimensional," she explained. "It's that person's PR for themselves." There's no substitute, she believes, for sitting down with a person one-on-one to get the full picture.

"When we meet our clients, we get a multifaceted impression," she said. "I may read on your profile that you love cats, but when I ask you about it, I learn that you had a beloved cat when you were three and now you're allergic to them. Or, I'll read a personality profile, but when I sit down with this person, I'll think, *Wow, I didn't know she had this kind of energy*. It wasn't reflected on the page."

While the Winkler clients fill out personality profiles similar to the ones found online, the difference, Ahlin believes, is the hour-and-a-half interview. Some of these matchmakers have a psychological background, but others are recruited for different reasons. "We go for people who have a heart, are good listeners, are empathetic, and who just have a feel for matching people for the long term," Ahlin told me. "On resumes, we look for evidence of good people skills—PR, customer service, nursing. It's not necessarily about an intellectual understanding. People either get it or they don't."

Ahlin estimates the agency's success rate at 70 percent— meaning that 70 percent of clients either end up in a relationship engineered by their matchmakers or get engaged to someone they've met through the agency. But unlike the studies being done at eHarmony, there's no follow-up to determine how long these relationships or marriages last or how satisfying they are down the line. Besides, Ahlin admitted, other variables may play a role in the high number of pairings. "When you pay 8 or 10 thousand dollars for a service like ours," she said, "you seriously want to find someone. It puts the notion 'I'm really ready' into your subconscious."

Ahlin and her matchmakers use feedback forms like those on Chemistry.com to learn how a match went after two clients have met in person. But whereas the Chemistry.com people classify this step as part of their scientific research, Ahlin says simply, "This way, you know what it is that works so you can get closer the next time—it helps us with intuition."

Often when Ahlin talks about intuition, she describes the same principles that the scientists I spoke with use in their empirically based matching systems. For instance, in matching couples, she follows what is essentially the similarity-complementarity model. "For a match to be successful," Ahlin said, "a couple's goals have to be the same; they have to want the same things in life." But, she added, "that doesn't mean they should be the same person. On the one hand, it's good if they have the same experiences, but sometimes having experiences that are different adds energy to the relationship."

Like Helen Fisher and Pepper Schwartz, Annie Ahlin believes that similarity and complementarity are situational models. "Each person is unique and contradictory," she told me, "and you can't just group people into big categories, the way the personality profiles do. So one person who is a Type A may be attracted to Type A in the beginning, but then we send them out and find out they need a Type B. So we adjust along the way. We're always adjusting. It's not a scientific process, it's an intuitive one."

Gian Gonzaga, the UCLA researcher hired by eHarmony, doesn't dismiss matchmakers. "I wouldn't be surprised if the basic constructs they're measuring are the exact same ones [that scientists measure]," he said. "Those who are good at matchmaking are the ones who get that four or five things are really critical."

I asked Gonzaga what those four or five things are, and he let out a long sigh.

"Oh, I don't know," he said, sheepishly. "It's funny enough, but I don't know. A similar sense of values. Other things, like agreeableness or warmth, are probably fairly important in terms of people matching up. You want two people who are relatively similar on wanting to cuddle or things like that."

At the word *cuddle,* I raised an eyebrow.

"It's kind of an unscientific term," he said, "but . . ."

I asked Gonzaga if using science to try to find lasting love might be too lofty a goal—a method that seems promising in theory but that turns out to be no more effective than consulting a matchmaker or cruising at your local bar. He disagreed.

"Imagine being in a bar," he said, "and how hard it would be to find five people you might connect with. If you actually match those people in the beginning, you're increasing your odds of meeting someone. Also, some people go to a bar to have a drink, some to meet people. We put people seriously looking for a relationship in one place, at the same time. So I think it's both the medium and it's the scale. And a matchmaker only knows so many people, but there are 8 million or 10 million users on eHarmony."

Moreover, in the future, science-based dating sites will evolve in ways that mimic real-world situations. Galen Buckwalter, eHarmony's research-and-development head, said that rather than relying on self-reports to assess how comfortable a person feels in social situations, his group is developing a model that will use computer simulation to immerse people in scenarios—a bar, a party, an intimate dinner—where variables like gender composition can be altered. "How does this person interact differently as the variables change?" Buckwalter asked. "I don't think we'll be relying on self-report 20 years from now. I think not only will data collection advance, but so will our analysis. We're just at the beginning, really."

Indeed, it may well take a generation before we learn whether the psychological, anthropological, or sociological model works best. Or maybe an entirely different theory will emerge. But at the very least, these dating sites and the relationships they spawn will help us to determine whether science has a place, and if so, how much of a place, in affairs of the heart.

Meanwhile, until these sites start sending me better dating prospects, I figured I'd take Neil Clark Warren up on his offer to introduce me to the 38-year-old single board member he thought would be such a good match for me. But when I asked a company spokesman about him, I was told that he had recently begun seeing someone. Did they meet through eHarmony? My potential soul mate declined to answer.

Jaron Lanier

Digital Maoism: The Hazards of the New Online Collectivism

The hive mind is for the most part stupid and bor-
ing. Why pay attention to it? The problem is in
the way the Wikipedia has come to be regarded
and used, how it's been elevated to such impor-
tance so quickly. And that is part of the larger pat-
tern of the appeal of a new online collectivism that
is nothing less than a resurgence of the idea that
the collective is all wise, that it is desirable to have
influence concentrated in a bottleneck that can
channel the collective with the most verity and
force. This is different from representative democ-
racy or meritocracy. This idea has had dreadful
consequences when thrust upon us from the
extreme Right or the extreme Left in various his-
torical periods. The fact that it's being reintro-
duced today by prominent technologists and futur-
ists, people who in many cases I know and like,
doesn't make it any less dangerous.

My Wikipedia entry identifies me (at least this week) as a
film director. It is true I made one experimental short film
about a decade and a half ago. The concept was awful: I
tried to imagine what Maya Deren would have done with

morphing. It was shown once at a film festival and was never distributed, and I would be most comfortable if no one ever sees it again.

In the real world it is easy to not direct films. I have attempted to retire from directing films in the alternative universe that is the Wikipedia a number of times, but somebody always overrules me. Every time my Wikipedia entry is corrected, within a day I'm turned into a film director again. I can think of no more suitable punishment than making these determined Wikipedia goblins actually watch my one small old movie.

Twice in the past several weeks, reporters have asked me about my filmmaking career. The fantasies of the goblins have entered that portion of the world that is attempting to remain real. I know I've gotten off easy. The errors in my Wikipedia bio have been (at least prior to the publication of this article) charming and even flattering.

Reading a Wikipedia entry is like reading the Bible closely. There are faint traces of the voices of various anonymous authors and editors, though it is impossible to be sure. In my particular case, it appears that the goblins are probably members or descendants of the rather sweet old *Mondo 2000* culture linking psychedelic experimentation with computers. They seem to place great importance on relating my ideas to those of the psychedelic luminaries of old (and in ways that I happen to find sloppy and incorrect.) Edits deviating from this set of odd ideas that are important to this one particular small subculture are immediately removed. This makes sense. Who else would volunteer to pay that much attention and do all that work?

The problem I am concerned with here is not the Wikipedia in itself. It's been criticized quite a lot, especially in the last year, but the Wikipedia is just one experiment that still has room to change and grow. At the very least it's a success at revealing what the online people with the most determination and time on their hands are thinking, and that's actually interesting information.

No, the problem is in the way the Wikipedia has come to be regarded and used, how it's been elevated to such importance so quickly. And that is part of the larger pattern of the appeal of a new online collectivism that is nothing less than a resurgence of the idea that the collective is all wise, that it is desirable to have influence concentrated in a bottleneck that can channel the collective with the most verity and force. This is different from representative democracy or meritocracy. This idea has had dreadful consequences when thrust upon us from the extreme Right or the extreme Left in various historical periods. The fact that it's now being reintroduced today by prominent technologists and futurists, people who in many cases I know and like, doesn't make it any less dangerous.

There was a well-publicized study in *Nature* last year comparing the accuracy of the Wikipedia to *Encyclopedia Britannica*. The results were a toss-up, while there is a lingering debate about the validity of the study. The items selected for the comparison were just the sort that Wikipedia would do well on: Science topics that the collective at large doesn't care much about. "Kinetic isotope effect" or "Vesalius, Andreas" are examples of topics that make the *Britannica* hard to maintain, because it takes work to find the right authors to research and review a multitude of diverse topics. But they are perfect for the Wikipedia. There is little controversy around these items, plus the Net provides ready access to a reasonably small number of competent specialist graduate student types possessing the manic motivation of youth.

A core belief of the wiki world is that whatever problems exist in the wiki will be incrementally corrected as the process unfolds. This is analogous to the claims of Hyper-Libertarians who put infinite faith in a free market or the Hyper-Lefties who are somehow able to sit through consensus decision-making processes. In all these cases, it seems to me that empirical evidence has yielded mixed results. Sometimes loosely structured collective activities yield continuous improvements, and sometimes they don't. Often we don't live long enough to find out. Later in this essay I'll point out what constraints make a collective smart. But first, it's important to not lose sight of values just because the question of whether a collective can be smart is so fascinating. Accuracy in a text is not enough. A desirable text is more than a collection of accurate references. It is also an expression of personality.

For instance, most of the technical or scientific information that is in the Wikipedia was already on the Web before the Wikipedia was started. You could always use Google or other search services to find information about items that are now wikified. In some cases I have noticed specific texts get cloned from original sites at universities or labs onto wiki pages. And when that happens, each text loses part of its value. Since search engines are now more likely to point you to the wikified versions, the Web has lost some of its flavor in casual use.

When you see the context in which something was written and you know who the author was beyond just a name, you learn so much more than when you find the same text placed in the anonymous, faux-authoritative, anticontextual brew of the Wikipedia. The question isn't just one of authentication and accountability, though those are important, but something more subtle. A voice should be sensed as a whole. You have to have a chance to sense personality in order for language to have its full meaning. Personal Web pages do that, as do journals and books. Even *Britannica* has an editorial voice, which some people have criticized as being vaguely too "Dead White Men."

If an ironic Web site devoted to destroying cinema claimed that I was a filmmaker, it would suddenly make sense. That would be an authentic piece of text. But placed out of context in the Wikipedia, it becomes drivel.

MySpace is another recent experiment that has become even more influential than the Wikipedia. Like the Wikipedia, it adds just a little to the powers already present on the Web in order to inspire a dramatic shift in use. MySpace is all about authorship, but it doesn't pretend to be all wise. You can always tell at least a little about the character of the person who made a MySpace page. But it is very rare indeed that a MySpace page inspires even the slightest confidence that the author is a trustworthy authority. Hurray for MySpace on that count!

MySpace is a richer, more multilayered, source of information than the Wikipedia, although the topics the two services cover barely overlap. If you want to research a TV show in terms of what people think of it, MySpace will reveal more to you than the analogous and enormous entries in the Wikipedia.

The Wikipedia is far from being the only online fetish site for foolish collectivism. There's a frantic race taking place online to become the most "Meta" site, to be the highest level aggregator, subsuming the identity of all other sites.

The race began innocently enough with the notion of creating directories of online destinations, such as the early incarnations of Yahoo. Then came AltaVista, where one could search using an inverted database of the content of the whole Web. Then came Google, which added page rank algorithms. Then came the blogs, which varied greatly in terms of quality and importance. This led to Metablogs such as Boing Boing, run by identified humans, which served to aggregate blogs. In all of these formulations, real people were still in charge. An individual or individuals were presenting a personality and taking responsibility.

These Web-based designs assumed that value would flow from people. It was still clear, in all such designs, that the Web was made of people and that ultimately value always came from connecting with real humans.

Even Google by itself (as it stands today) isn't Meta enough to be a problem. One layer of page ranking is hardly a threat to authorship, but an accumulation of many layers can create a meaningless murk, and that is another matter.

In the last year or two the trend has been to remove the scent of people, so as to come as close as possible to simulating the appearance of content emerging out of the Web as if it were speaking to us as a supernatural oracle. This is where the use of the Internet crosses the line into delusion.

Kevin Kelly, the former editor of *Whole Earth Review* and the founding executive editor of *Wired,* is a friend and someone who has been thinking about what he and others call the "hive mind." He runs a Web site called Cool Tools that's a cross between a blog and the old *Whole Earth Catalog.* On Cool Tools, the contributors, including me, are not a hive because we are identified.

In March, Kelly reviewed a variety of "Consensus Web filters" such as Digg and Reddit that assemble material every day from all the myriad of other aggregating sites. Such sites intend to be more Meta than the sites they aggregate. There is no person taking responsibility for what appears on them, only an algorithm. The hope seems to be that the most Meta site will become the mother of all bottlenecks and receive infinite funding.

That new magnitude of Metaness lasted only a month.

In April, Kelly reviewed a site called "popurls" that aggregates consensus Web filtering sites . . . and there was a new "most Meta." We now are reading what a collectivity algorithm derives from what other collectivity algorithms derived from what collectives chose from what a population of mostly amateur writers wrote anonymously.

Is popurls any good? I am writing this on May 27, 2006. In the last few days an experimental approach to diabetes management has been announced that might prevent nerve damage. That's huge news for tens of millions of Americans. It is not mentioned on popurls. Popurls does clue us in to this news: "Student sets simultaneous world ice cream-eating record, worst ever ice cream headache." Mainstream news sources all lead today with a serious earthquake in Java. Popurls includes a few mentions of the event, but they are buried within the aggregation of aggregate news sites like Google News. The reason the quake appears on popurls at all can be discovered only if you dig through all the aggregating layers to find the original sources, which are those rare entries actually created by professional writers and editors who sign their names. But at the layer of popurls, the ice cream story and the Javanese earthquake are at best equals, without context or authorship.

Kevin Kelly says of the popurls site, "There's no better way to watch the hive mind." But the hive mind is for the most part stupid and boring. Why pay attention to it?

Readers of my previous rants will notice a parallel between my discomfort with so-called Artificial Intelligence and the race to erase personality and be most Meta. In each case, there's a presumption that something like a distinct kin to individual human intelligence is either about to appear any minute or has already appeared. The problem with that presumption is that people are all too willing to lower standards in order to make the purported newcomer appear smart. Just as people are willing to bend over backward and make themselves stupid in order to make an AI interface appear smart (as happens when someone can interact with the notorious Microsoft paper clip), so are they willing to become uncritical and dim in order to make Meta-aggregator sites appear to be coherent.

There is a pedagogical connection between the culture of Artificial Intelligence and the strange allure of anonymous collectivism online. Google's vast servers and the Wikipedia are both mentioned frequently as being the startup memory for Artificial Intelligences to come. Larry Page is quoted via a link presented to me by popurls this morning (who knows if it's accurate) as speculating that an AI might appear within Google within a few years. George Dyson has wondered if such an entity already exists on the Net, perhaps perched within Google. My point here is not to argue about the existence of Metaphysical entities but just to emphasize how premature and dangerous it is to lower the expectations we hold for individual human intellects.

The beauty of the Internet is that it connects people. The value is in the other people. If we start to believe that the Internet itself is an entity that has something to say, we're devaluing those people and making ourselves into idiots.

Compounding the problem is that new business models for people who think and write have not appeared as quickly as we all hoped. Newspapers, for instance, are on the whole facing a grim decline as the Internet takes over the feeding of curious eyes that hover over morning coffee and, even worse, classified ads. In the new environment, Google News is for the moment better funded and enjoys a more secure future than most of the rather small number of fine reporters around the world who ultimately create most of its content. The aggregator is richer than the aggregated.

The question of new business models for content creators on the Internet is a profound and difficult topic in itself, but it must at least be pointed out that writing professionally and well takes time and that most authors need to be paid to take that time. In this regard, blogging is not writing. For example, it's easy to be loved as a blogger. All you have to do is play to the crowd. Or you can flame the crowd to get attention. Nothing is wrong with either of those activities. What I think of as real writing, however, writing meant to last, is something else. It involves articulating a perspective that is not just reactive to yesterday's moves in a conversation.

The artificial elevation of all things Meta is not confined to online culture. It is having a profound influence on how decisions are made in America.

What we are witnessing today is the alarming rise of the fallacy of the infallible collective. Numerous elite organizations have been swept off their feet by the idea. They are inspired by the rise of the Wikipedia, by the wealth of Google, and by the rush of entrepreneurs to be the most Meta. Government agencies, top corporate planning departments, and major universities have all gotten the bug.

As a consultant, I used to be asked to test an idea or propose a new one to solve a problem. In the last couple of years I've often been asked to work quite differently. You might find me and the other consultants filling out survey forms or tweaking edits to a collective essay. I'm saying and doing much less than I used to, even though I'm still being paid the same amount. Maybe I shouldn't complain, but the actions of big institutions do matter, and it's time to speak out against the collectivity fad that is upon us.

It's not hard to see why the fallacy of collectivism has become so popular in big organizations: If the principle is correct, then individuals should not be required to take on risks or responsibilities. We live in times of tremendous uncertainties coupled with infinite liability phobia, and we must function within institutions that are loyal to no executive, much less to any lower level member. Every individual who is afraid to say the wrong thing within his or her organization is safer when hiding behind a wiki or some other Meta-aggregation ritual.

I've participated in a number of elite, well-paid wikis and Metasurveys lately and have had a chance to observe the results. I have even been part of a wiki about wikis. What I've seen is a loss of insight and subtlety, a disregard for the nuances of considered opinions, and an increased tendency to enshrine the official or normative beliefs of an organization. Why isn't everyone screaming about the recent epidemic of inappropriate uses of the collective? It seems to me the reason is that bad old ideas look confusingly fresh when they are packaged as technology.

The collective rises around us in multifarious ways. What afflicts big institutions also afflicts pop culture. For instance, it has become notoriously difficult to introduce a new pop star in the music business. Even the most successful entrants have hardly ever made it past the first album in the last decade or so. The exception is *American Idol*. As with the Wikipedia, there's nothing wrong with it. The problem is its centrality.

More people appear to vote in this pop competition than in presidential elections, and one reason for this is the instant convenience of information technology. The collective can vote by phone or by texting, and some vote more than once.

The collective is flattered, and it responds. The winners are likable, almost by definition.

But John Lennon wouldn't have won. He wouldn't have made it to the finals. Or if he had, he would have ended up a different sort of person and artist. The same could be said about Jimi Hendrix, Elvis, Joni Mitchell, Duke Ellington, David Byrne, Grandmaster Flash, Bob Dylan (please!), and almost anyone else who has been vastly influential in creating pop music.

As below, so above. The *New York Times,* of all places, has recently published Op-Ed pieces supporting the pseudoidea of intelligent design. This is astonishing. The *Times* has become the paper of averaging opinions. Something is lost when *American Idol* becomes a leader instead of a follower of pop music. But when intelligent design shares the stage with real science in the paper of record, everything is lost.

How could the *Times* have fallen so far? I don't know, but I would imagine the process was similar to what I've seen in the consulting world of late. It's safer to be the aggregator of the collective. You get to include all sorts of material without committing to anything. You can be superficially interesting without having to worry about the possibility of being wrong.

Except when intelligent thought really matters. In that case the average idea can be quite wrong, and only the best ideas have lasting value. Science is like that.

The collective isn't always stupid. In some special cases the collective can be brilliant. For instance, there's a demonstrative ritual often presented to incoming students at business schools. In one version of the ritual, a large jar of jelly beans is placed in the front of a classroom. Each student guesses how many beans there are. While the guesses vary widely, the average is usually accurate to an uncanny degree.

This is an example of the special kind of intelligence offered by a collective. It is that peculiar trait that has been celebrated as the "Wisdom of Crowds," though I think the word *wisdom* is misleading. It is part of what makes Adam Smith's Invisible Hand clever and is connected to the reasons Google's page rank algorithms work. It was long ago adapted to futurism, where it was known as the Delphi technique. The phenomenon is real and immensely useful.

But it is not infinitely useful. The collective can be stupid, too. Witness tulip crazes and stock bubbles. Hysteria over fictitious satanic cult child abductions. Y2K mania.

The reason the collective can be valuable is precisely that its peaks of intelligence and stupidity are not the same as the ones usually displayed by individuals. Both kinds of intelligence are essential.

What makes a market work, for instance, is the marriage of collective and individual intelligence. A marketplace can't exist only on the basis of having prices determined by competition. It also needs entrepreneurs to come up with the products that are competing in the first place.

In other words, clever individuals, the heroes of the marketplace, ask the questions that are answered by collective behavior. They put the jelly beans in the jar.

There are certain types of answers that ought not be provided by an individual. When a government bureaucrat sets a price, for instance, the result is often inferior to the answer that would come from a reasonably informed collective that is reasonably free of manipulation or runaway internal resonances. But when a collective designs a product, you get design by committee, which is a derogatory expression for a reason.

Here I must take a moment to comment on Linux and similar efforts. The various formulations of "open" or "free" software are different from the Wikipedia and the race to be most Meta in important ways. Linux programmers are not anonymous, and in fact personal glory is part of the motivational engine that keeps such enterprises in motion. But there are similarities, and the lack of a coherent voice or design sensibility in an aesthetic sense is one negative quality of both open source software and the Wikipedia.

These movements are at their most efficient while building hidden information plumbing layers, such as Web servers. They are hopeless when it comes to producing fine user interfaces or user experiences. If the code that ran the Wikipedia user interface were as open as the contents of the entries, it would churn itself into impenetrable muck almost immediately. The collective is good at solving problems that demand results that can be evaluated by uncontroversial performance parameters, but it is bad when taste and judgment matter.

Collectives can be just as stupid as any individual, and in important cases, stupider. The interesting question is whether it's possible to map out where the one is smarter than the many.

There is a lot of history to this topic, and varied disciplines have lots to say. Here is a quick pass at where I think the boundary between effective collective thought and nonsense lies: The collective is more likely to be smart when it isn't defining its own questions, when the goodness of an answer can be evaluated by a simple result (such as a single numeric value), and when the information system that informs the collective is filtered by a quality control mechanism that relies on individuals to a high degree. Under those circumstances, a collective can be smarter than a person. Break any one of those conditions and the collective becomes unreliable or worse.

Meanwhile, an individual best achieves optimal stupidity on those rare occasions when one is both given substantial powers and insulated from the results of his or her actions.

If the above criteria have any merit, then there is an unfortunate convergence. The setup for the most stupid collective is also the setup for the most stupid individuals.

Every authentic example of collective intelligence that I am aware of also shows how that collective was guided or inspired by well-meaning individuals. These people focused the collective and in some cases also corrected for some of the common hive mind failure modes. The balancing of influence between people and collectives is the heart of the design of democracies, scientific communities, and many other long-standing projects. There's a lot of experience out there to work with. A few of these old ideas provide interesting new ways to approach the question of how to best use the hive mind.

The pre-Internet world provides some great examples of how personality-based quality control can improve collective intelligence. For instance, an independent press provides tasty news about politicians by reporters with strong voices and reputations, like the Watergate reporting of Woodward and Bernstein. Other writers provide product reviews, such as Walt Mossberg in the *Wall Street Journal* and David Pogue in the *New York Times*. Such journalists inform the collective's determination of election results and pricing. Without an independent press, composed of heroic voices, the collective becomes stupid and unreliable, as has been demonstrated in many historical instances. (Recent events in America have reflected the weakening of the press, in my opinion.)

Scientific communities likewise achieve quality through a cooperative process that includes checks and balances and ultimately rests on a foundation of goodwill and "blind" elitism—blind in the sense that ideally anyone can gain entry but only on the basis of a meritocracy. The tenure system and many other aspects of the academy are designed to support the idea that individual scholars matter, not just the process or the collective.

Another example: Entrepreneurs aren't the only "heroes" of a marketplace. The role of a central bank in an economy is not the same as that of a Communist Party official in a centrally planned economy. Even though setting an interest rate sounds like the answering of a question, it is really more like the asking of a question. The Fed asks the market to answer the question of how to best optimize for lowering inflation, for instance. While that might not be the question everyone would want to have asked, it is at least coherent.

Yes, there have been plenty of scandals in government, in the academy, and in the press. No mechanism is perfect, but still here we are, having benefited from all of these institutions. There certainly have been plenty of bad reporters, self-deluded academic scientists, incompetent bureaucrats, and so on. Can the hive mind help keep them in check? The answer provided by experiments in the pre-Internet world is "yes" but only provided some signal processing is placed in the loop.

Some of the regulating mechanisms for collectives that have been most successful in the pre-Internet world can be understood in part as modulating the time domain. For instance, what if a collective moves too readily and quickly, jittering instead of settling down to provide a single answer?

This happens on the most active Wikipedia entries, for example, and has also been seen in some speculation frenzies in open markets.

One service performed by representative democracy is low-pass filtering. Imagine the jittery shifts that would take place if a wiki were put in charge of writing laws. It's a terrifying thing to consider. Superenergized people would be struggling to shift the wording of the tax code on a frantic, never-ending basis. The Internet would be swamped.

Such chaos can be avoided in the same way it already is, albeit imperfectly, by the slower processes of elections and court proceedings. The calming effect of orderly democracy achieves more than just the smoothing out of peripatetic struggles for consensus. It also reduces the potential for the collective to suddenly jump into an overexcited state when too many rapid changes to answers coincide in such a way that they don't cancel each other out. (Technical readers will recognize familiar principles in signal processing.)

The Wikipedia has recently slapped a crude low-pass filter on the jitteriest entries, such as "President George W. Bush." There's now a limit to how often a particular person can remove someone else's text fragments. I suspect that this will eventually have to evolve into an approximate mirror of democracy as it was before the Internet arrived.

The reverse problem can also appear. The hive mind can be on the right track but moving too slowly. Sometimes collectives would yield brilliant results given enough time, but there isn't enough time. A problem like global warming would automatically be addressed eventually if the market had enough time to respond to it, for instance. Insurance rates would climb, and so on. Alas, in this case there isn't enough time, because the market conversation is slowed down by the legacy effect of existing investments. Therefore some other process has to intervene, such as politics invoked by individuals.

Another example of the slow hive problem: There was a lot of technology developed slowly in the millennia before there was a clear idea of how to be empirical and of how to have a peer reviewed technical literature and an education based on it and before there was an efficient market to determine the value of inventions. What is crucial to notice about modernity is that structure and constraints were part of what sped up the process of technological development, not just pure openness and concessions to the collective.

Let's suppose that the Wikipedia will indeed become better in some ways, as is claimed by the faithful, over a period of time. We might still need something better sooner.

Some wikitopians explicitly hope to see education subsumed by wikis. It is at least possible that in the fairly near future enough communication and education will take place through anonymous Internet aggregation that we could become vulnerable to a sudden dangerous empowering of the hive mind. History has shown us again and again that a hive mind is a cruel idiot when it runs on autopilot. Nasty hive mind outbursts have been flavored Maoist, Fascist, and religious, and these are only a small sampling. I don't see why there couldn't be future social disasters that appear suddenly under the cover of technological utopianism. If wikis are to gain any more influence they ought to be improved by mechanisms like the ones that have worked tolerably well in the pre-Internet world.

The hive mind should be thought of as a tool. Empowering the collective does not empower individuals—just the reverse is true. There can be useful feedback loops set up between individuals and the hive mind, but the hive mind is too chaotic to be fed back into itself.

These are just a few ideas about how to train a potentially dangerous collective and not let it get out of the yard. When there's a problem, you want it to bark but not bite you.

The illusion that what we already have is close to good enough, or that it is alive and will fix itself, is the most dangerous illusion of all. By avoiding that nonsense, it ought to be possible to find a humanistic and practical way to maximize value of the collective on the Web without turning ourselves into idiots. The best guiding principle is to always cherish individuals first.

It Should Happen to You

The anxieties of YouTube fame

Stevie Ryan received her first Oscar, after a fashion, this year, at the age of 22, only 18 months after moving to Los Angeles to become a movie star. She grew up in California's high desert, a couple of hours to the east, in a town along the road to Las Vegas called Victorville. Her parents worked at calibrating truck scales for weigh stations on the interstate—a family business going back two generations on her mom's side. Throughout her childhood and adolescence, Ryan harbored escape fantasies involving the Hollywood of her parents' and grandparents' generations—Lucille Ball, Audrey Hepburn, Buster Keaton, Clara Bow—but she never participated in high-school theatrical productions. She did attend her high-school prom dressed as Marilyn Monroe, down to the elbow-length gloves. (Her date wore a Mohawk and muttonchops.) After a brief stint in community college, she concluded that she was "too right-brain for school" and followed an older brother to Huntington Beach—anything to get out of Victorville. "Then I thought, Screw these people—I'll just go to LA, see what happens," she said recently.

The Oscar was delivered rather unceremoniously—not

in March, at the Academy Awards, but in August, three and a half minutes into a sketch Ryan was filming, while she was still in character as Cynthia, an 18-year-old Latina from East LA who is better known as Little Loca, after the handle Ryan uses when she uploads some of her homemade sketches onto the video-sharing site YouTube. This was about the 40th in a series of short Little Loca videos that had by then attracted over a million viewings, thanks to Loca's "big old mouth" (both literally—her heavily outlined lips command attention—and figuratively) and her irreverent putdowns ("You better watch out, *fool,* because God's gonna come around and strike you down with some *lightning* if you don't be *careful*"). Loca was wearing a bandanna and hoop earrings and sitting on a sofa, against a plain white wall, between two women who were known to regular viewers as Smiley (a friend of Ryan's) and Silent Girl (Ryan's cousin). Rap music was playing in the background.

"*Damn,* this shit is *heavy,*" Loca said, in a pronounced Hispanic accent, after accepting the gold statuette from Smiley and waving it around. "I could knock somebody *out* with this." Then she launched into an earnest acceptance speech. "I want to thank YouTube," she said. "You're so important in my life right now. And without YouTube there's no way in *hell* Loca could have, you know, got something like this."

It seemed to be a genuine Oscar—stolen from a bar by a friend of Ryan's—and the moment was rich with postmodern significance. Over the previous three months, Loca's fans, many of them Hispanic, had warmed to her story: spunky ghetto kid—a *chola*—with an overprotective older brother, a 4.0 grade-point average, and her innocence proudly intact. (That gang sign that she seemed to flash at the end of each video was really a sideways *V,* for virgin.) They knew she'd been prom queen, and they had met her onetime boyfriend Raúl. They'd learned that Silent Girl went mute after the death of her brother, an innocent bystander in a botched robbery. And they'd grown accustomed to Loca's distinctive, almost bewitching screen presence—the way her dark eyebrows and pursed lips slide effortlessly from a knowing smile to an outraged glare. At the same time, they'd begun noticing suspicious details that called into question the diary's authenticity: the mole on Loca's right cheek seemed to vary in size and placement; Raúl bore a striking resemblance to Drake Bell, the costar of Nickelodeon's *Drake and Josh,* a teen sitcom; and didn't Loca resemble a young woman—a white woman—named Stevie Ryan, who'd been photographed with Drake Bell at the MTV Movie Awards, in June? Accepting the Oscar as Loca was Stevie Ryan's tacit way of acknowledging the act while also congratulating herself on having legitimately achieved a kind of alternate-reality stardom. Smiley and Silent Girl wore black Little Loca T-shirts they'd bought on the Web from a total stranger.

Loca's outing mirrored, in some ways, that of the season's most famous Internet adolescent, LonelyGirl15, whose homespun, if sharply edited, tales of science projects, boy troubles, and religion captivated millions of YouTube viewers before she was exposed as the creation of filmmakers represented by the Creative Artists Agency on Wilshire Boulevard, instead of, say, a girl in her bedroom on some sleepy midwestern Main Street. But whereas the people behind LonelyGirl15 were interested, from the outset, in exploring the possibilities of a "new art form," as they called it, unfolding in two-minute episodes, Stevie Ryan came by her YouTube celebrity accidentally, while killing time between auditions and acting classes.

Ryan's show-business career started when she landed a bit part in a Hilary Duff video (playing Marilyn Monroe) as a result of her first audition, while still living in Victorville. That was all the encouragement she needed, and before long she was dating Bell, whom she met in Huntington Beach. But steady work proved hard to come by, and her reel, after more than a year in LA, was a typically mixed bag: another music gig (a Billy Idol video), a Japanese commercial, modeling for a fashion start-up. She got a job working at a Levi's store in Beverly Hills.

Six months ago, she borrowed Bell's Sony Handycam and started making videos. They were mostly vintage-style silent films, with names like *Beyond the Sea . . .* and *Satin Doll,* which she edited, with no formal training, using Windows Movie Maker. She experimented with uploading a few of the films onto YouTube and only then discovered the site's ruthlessly populist ethos: what people seemed to like was not pretentious art films with obvious Hollywood aspirations but the confessional blogs of young girls in their bedrooms. Little Loca—a composite of the tough-talking, strong-willed *cholas* Ryan used to admire in Victorville—was born.

Within a few weeks, YouTube became a full-time pursuit for Ryan. "It's basically all I do," she told me. In addition to Loca, she began doing spoofs and impressions of established YouTube bloggers (a surefire way of getting attention) and kept up, sporadically, with the artsy silent films. The quest for stardom that had led her to Hollywood now pitted her against nonprofessionals in Toronto and Pittsburgh and Tasmania.

Three days after Little Loca's Oscar speech, a 79-year-old widower named Peter turned on his Webcam, in the English countryside, and announced, "I got addicted to YouTube. It's a fascinating place to go to see all the wonderful videos that you young people have produced, so I thought I'd have a go at doing one myself." (About half of all registered YouTube users are said to be under 20.) He was wearing a beige V-neck sweater and glasses and sat in front of 1970s-era wallpaper and a small painting of a motorcycle. "Oh, yes, and, incidentally, I really am as old as I look," he said. "What I hope I'll be able to do is just bitch and grumble about life in general from the perspective of an old person who's been there and done that."

Peter called himself geriatric1927, after the year of his birth, and uploaded the video, which was two minutes long, under the title *First Try*. It had been viewed scarcely more than 300 times when it came to the attention of a staffer at YouTube headquarters, in San Mateo, California, who showed it to Maryrose Dunton, YouTube's director of product management. She is one of the people in charge of selecting videos to feature on the YouTube home page, which serves as an informal recommendation list. Of the 70,000 videos added to the site every day, fewer than a dozen receive this special treatment. Dunton, who says she is "totally fascinated by old people and tech," put Peter's video at the top of the featured list. The YouTube audience, bombarded by frenetic, attention-seeking teens, immediately warmed to Peter's reserve. By the following week, geriatric1927, who had begun narrating his life story, from primary school through the Blitz and on into health-department work in Leicestershire, without ever leaving his chair, had more subscribers than any other user in YouTube's history. *First Try* has now been seen nearly 2 million times.

One hesitates to cite these statistics, because the story of YouTube, since its launch, 10 months ago, has been one of exponential growth, at times challenging the company's abilities to cope with the demand on its servers. (Bandwidth costs are thought to exceed a million dollars a month.) Last week, according to Alexa, a Web-traffic monitor, it was the

10th-biggest site on the Internet, drawing more visits than eBay, Amazon, or Wikipedia. By late summer, there were approximately 6 million videos archived on the site, and daily viewings had crossed the 100 million mark, a great many of them devoted not to original content, such as Peter's or Stevie Ryan's, but to preexisting footage in a wide range of genres: weird home movies (an old woman punching another old woman in the face), sports (Zinedine Zidane's infamous head butt), music (Hendrix playing "The Star-Spangled Banner"), and politics (Senator George Allen referring to a rival's campaign worker as *macaca;* Bill Clinton attacking Fox News on Fox News).

YouTube was founded in February 2005, in a Silicon Valley garage, by a couple of former PayPal employees, Steve Chen and Chad Hurley. Their background was technological, not visionary. They aimed to provide an easy interface for storing, sorting, and sharing the kinds of digital videos that, thanks to cell-phone cameras and Webcams, have become more and more prevalent. When, in late August, I visited the YouTube offices, which sit above a pizza parlor on the main commercial strip in downtown San Mateo, several of the 60 or so employees had just finished watching clips of a dance number from the previous night's Emmy Awards show, in which the host, Conan O'Brien, sang, "At this very moment your kids are on YouTube watching a cat on a toilet." Julie Supan, YouTube's senior director of marketing, handed me a copy of a recent *People Hollywood Daily*. Its cover read, "Television's Brave New World: How the YouTube Revolution Is Changing Everything You Knew about the Industry." She was unclear about what, specifically, the YouTube revolution is, however. "We don't have time to stop and think a lot," she said.

Hurley, the company's CEO, told me that he wanted to "democratize the entertainment process," but YouTube's business model remains somewhat undefined. The found footage that generates the bulk of its traffic is, in many cases, subject to copyright restrictions, leaving YouTube vulnerable to lawsuits. ("The only reason it hasn't been sued yet is because there is nobody with big money to sue," Mark Cuban, the cofounder of HDNet, said recently.) Networks like NBC and Fox have intervened to request that particular clips—"Lazy Sunday," from *Saturday Night Live,* or Clinton's Fox appearance—be taken down. (Fox later relented, possibly because of complaints of censorship; NBC has begun uploading promotional spots, if not actual footage.)

YouTube's long-term strength seems to lie in the devoted community of users and bloggers (or "broadcasters," as the company likes to call them), some of whom turn out to have crossover potential. Brooke Brodack, a skinny, gap-toothed, 20-year-old receptionist from western Massachusetts, became, in effect, the first real YouTube star when she was hired in June by Carson Daly to develop content for his production company on the basis of her defiantly madcap skits and lip-synching.

"They want to be seen, and we're providing the largest audience for that," Hurley said. "But I think the stars on the site don't necessarily translate to television." His plan is to develop a new advertising model that's "not forced on the user." Yet the site's popularity stems from its openness—anyone can upload a video—which makes much of the content difficult to monitor and target ads for. Hurley has therefore begun experimenting with "branded channels," and he pointed to the recently launched Paris Hilton channel as an example. In a joint arrangement with Warner Bros., Hilton's record label, and Fox, which sponsored her channel to promote one of its new shows, her videos—Paris waving at fans in Tokyo, Paris having her hair done—

received front-page placement, just like the featured spots. YouTube has also agreed to provide the Warner Music Group with "fingerprinting" technology that will help locate its copyrighted material on the site, which it will be free to authorize or remove as it chooses. Warner will upload its own music-video library and will share the revenues from advertising targeted at its content.

Perhaps the best case for YouTube as a "democratizer" is Peter the geriatric. "What's interesting to me is he doesn't really have a different story," Maryrose Dunton said. "He wasn't famous. He's just this average old guy, like, telling his story. That's so endearing."

But geriatric1927 was not, in an important sense, a truly democratic star. Like an aspiring model who is spotted in a drugstore by a hot-shot agent, he'd been plucked from the crowd and thrust directly into the spotlight. Ernie Rogers, a 23-year-old guitar player in San Bernardino, may represent the ultimate realization—and corruption—of YouTube's democratic ideal. Although on his user profile he bills himself as a "typical guy," Rogers, who goes by the name lamo1234, has watched more than 900,000 videos on YouTube since May. That averages out to approximately 250 per hour, not allowing for sleep. What he watches, primarily, are his own guitar solos (or the first few seconds of them), over and over, to boost his view counts to levels that will make others take notice. His strategy seems to have been successful: one of his solos, a medley of Nirvana, Guns N' Roses, and Beethoven licks, has been viewed 200,000 times—and only 60,000 of those viewings were by him. Unfortunately, this strategy leaves little time for actually playing music. "Next year, the No. 1 spot on YouTube is going to be me, every day," he told me. "I just need to make my band."

Stevie Ryan has a pale, egg-shaped face and dark-red hair that she likes to run her hands through when she's not waving them about—punctuation for the many occasions she finds to use the words *cool* and *awesome*. She shares an apartment just off Melrose Avenue with Kendal Sheppard, a young woman she met in an acting class. The apartment, which is just a few blocks from the country's last silent-movie theater, is decorated with memorabilia honoring Ryan's real-world idols—posters of Marilyn Monroe and James Dean, an *I Love Lucy* pillow—but when I visited, not long ago, she seemed most animated while discussing her Internet rivals, people with names like LisaNova ("Don't mention her in this article, 'cause I don't want her to get attention") and Vvvvalentine ("Oh, I love this girl—and her videos are about absolutely nothing!") and FilthyWhore ("She's a fucking bitch"). LonelyGirl15 had not yet been outed, but Ryan was already certain that her YouTube blogs were the work of an actress. "Oh, my God, she's the worst deliverer of lines *ever,*" Ryan said, cuing up a LonelyGirl episode on her laptop and fast-forwarding to a spot where Bree, as the protagonist is called, claps her hands over her ears and says, without blinking, "My parents are unable to see things from my point of view no matter how much I try and explain it to them." Ryan shook her head. "That, to me, is just *so* fake," she said. "I don't understand it."

Ryan's Little Loca videos are also fake, of course, but a number of plot points have been drawn from Ryan's own life. For instance, Kendal Sheppard has a Cairn terrier named Baxter, so, rather than risk having unexplained barking in the background, Ryan wrote Baxter into the script, as a stray that Loca found one day while jogging. In July, back in Victorville, which Ryan calls "the meth capital of the United States," her car was broken into and vandalized. Her first thought was: It's a very Loca situation. So she got into character, put on the accent, and shot an episode inside the car, amid the wreckage. In some cases, the distinction seems to have blurred almost to the point of identity confusion. At one moment in the car-theft video, Ryan/Loca, who is visibly upset, says to the camera, "And this is real, you guys. I'm not trying to play no stupid YouTube joke or nothing. . . . I feel so *invaded,* and I just feel like everybody's *watching* me, you know?"

"Loca is single—or Stevie, whatever people call me," Ryan said to me, explaining that she and Drake Bell had split up earlier in the summer. Bell's disappearance from Ryan's life necessitated a falling out between Loca and Raúl as well. Ryan was clearly having cathartic fun with the exercise, blaming Raúl for the car incident (he needed cash to buy drugs) and, in another episode, mentioning that he'd been beaten up. ("Homegirl was telling me that, you know, they were *hitting* him and *kicking* him up on the floor and stuff like that.")

During a recent trip to San Francisco, Ryan told me, she had been accosted by a group of teens at a mall, wanting to know if she was "Little Loca from YouTube." (She was upset that she'd been caught off guard and hadn't looked her best for all the pictures they snapped.) She also, thanks to Loca, was now being represented by a Hollywood agency. "Seriously, if you Googled me, like, a couple months ago, you wouldn't get crap," she said, typing her name into the search engine. "I'm just a normal person. And now you actually get stuff. It's, like, crazy. That's more than I could ever ask for, just to be on Google." The search led to a fan site for various celebrities; Stevie Ryan's name and head shot were featured alongside Tom Cruise, Rachel McAdams, and Johnny Depp.

Along with the arrival of a Google track record had

come some anxieties about her place in the Hollywood pecking order, where, the revolution notwithstanding, YouTube
still doesn't count for much. She'd been embarrassed at a
recent party when she wondered whether other guests were
being patronizing about her Little Loca pursuits. "As weird
as it sounds, being in LA, with all these actors, nobody wants
to do it," she said. "There's this whole thing in LA where, if
it's not on a billboard, it's not really acting." She'd been trying to persuade Sheppard, who once starred in the MTV
reality series Road Rules, another ghettoized genre, to make
videos with her, to no avail.

Ryan flipped through her high-school yearbook and volunteered, "You can see I was a lot fatter than I am now." (Her
Google search bar had recognized the name Stevie Ryan and
suggested two related searches that she had entered previously: "Stevie Ryan thin" and "Stevie Ryan skinny.")

Another anxiety grew out of a suspicion that YouTube
was screwing her over, artificially suppressing her page
views and going out of its way not to "feature" her the way
it had featured geriatric1927. "OK, seriously? They do not
like me on here," she said. "They hate my guts. I've never
been featured, so I don't watch the featured videos now. I'm
really angry at YouTube. I don't care what anybody says;
they're doing it on purpose. I have written probably like, I
don't know, a million letters."

The transformation from Stevie Ryan to Little Loca takes
about 15 minutes and requires both a minor makeover
(drawing the mole, teasing her hair, applying brown lipstick) and a personality adjustment, giving a strident edge to
Ryan's blithe Valley Girl persona. (*Cool* and *awesome*
become *scandalous* and *nasty*.) Stevie wears heels; Loca wears
Nike Cortez sneakers, big hoop earrings, and a cross around
her neck.

Ryan retrieved a marble composition book in which she'd jotted some notes for her next video, not so much a script as an outline. One item read, "Paris Hilton and her big old nasty feet"—a reminder to talk trash about YouTube's latest interloper. Another said, "Dog shots!" ("I really did take him the other day to get him shots, and I wanted to talk about why that shit really did cost me $150," she said.) There was also a plot-moving device: "Silent Girl is mad because I've been talking to Raúl."

Ryan said that she prefers to shoot Little Loca videos straight through, without editing, to create the genuine feel of a video blog. She tends to ramble, however, and her videos, which once averaged about three minutes, now run for six or seven. ("Loca has too much to say.") Before starting, she read over the viewer-feedback comments posted below her latest video, like a kind of pep talk. "People are so mean on here," she said, after reading one particularly stinging insult, by a fellow-broadcaster named Mojojojoe69. "I give this guy all these shout-outs all the time."

She sat on a bright-red sofa and held the camera in her right hand, just in front of her face. "*Hey,* everybody, what up? It's your *homegirl,* Loca, all up in the *house,*" she started, before tearing into her critics. "All I see is you fools over here, crying like a bunch of damn *babies,* including *you,* Mojojojoe69. What the *hell* is *your* problem? . . . Either you better *respect* or you better get the *hell* out of here, fool. Make your *damn* mind up, all right? You can't talk shit to Loca one second and be her *homey.*" After a minute or so, Ryan stopped, concluding that she was talking too fast. Take two ran for four minutes before she stopped again. ("I don't remember if I talked about Mojojojoe.") The third try, despite faltering arm strength, stuck: six minutes and 33 seconds. She downloaded it onto her laptop and then, opening Windows Movie Maker, selected "grayscale," converting it to a highly saturated (for YouTube) black-and-white. Next, she searched her music library for an old-school rap track ("Loca's really gangsta," she said), ultimately selecting "Protect Ya Neck," by the Wu-Tang Clan, which she set to play on a separate track underneath her voice. A series of selections—"fade," "in and out," "moving titles," "layered"— created a template for the opening credits. She selected a graffiti font and gave the video a title: *The Locamotion Isn't Happy with You Guys.*

"I seriously can't sit through my videos," she said, after uploading it onto the site. "They're, like, so annoying." She combed out her hair; removed her makeup; and then got in her car and drove to the actor Crispin Glover's house, in Silver Lake, to discuss plans for possible joint projects: a YouTube star landing a feature-film role and a Hollywood star joining the YouTube community.

Crispin Glover was wearing a black jacket, black pants, black shoes, and black socks. His front door was open, and in the yard were a number of antique cars covered with tarps. Reverberation from an indie-rock performance, part of a street festival on Sunset Boulevard, made its way up the hill. Ryan curled up in a chair in the living room and looked adoringly at Glover.

A few weeks earlier, Glover had sent her a fan letter through MySpace (its subject was "Genuine Crispin Glover writes Stevie Ryan"). It said that he was planning to make a "party film," set at a castle he owned in the Czech Republic, and he was looking for talented (and inexpensive) actors to work with.

"I saw that she was doing these things that fooled people and that she had these various characters," Glover said, explaining his initial attraction to her YouTube work. "And I've had some experience with that as well—things that I've done people will think different things about what they actually are. I like that."

A search for Glover's name on YouTube will uncover, among other perplexing things, at least 10 different uploads of a 1987 appearance on David Letterman's show. Glover comes onstage wearing a wig, funny glasses, tight striped pants, and platform shoes. He behaves very strangely (one of the YouTube clips was uploaded with the title *Drugs Are Baaaad*); Letterman is not amused. Glover challenges Letterman to an arm-wrestling contest and then karate-kicks in the direction of Letterman's head. Letterman rises from his desk and walks offstage. YouTube users have watched the incident more than a quarter million times. "It's interesting now that there's this whole new life for it," Glover said.

Glover spent much of the past decade writing, directing, and producing a film called *What Is It?* using mostly actors with Down syndrome, which he plans to personally present at art theaters in selected cities this fall. (He himself plays a character called Dueling Demi-God Auteur.) He didn't seem to have much interest in exploring YouTube's potential as a new narrative format, to compete with television and feature films, and instead saw it as a forum for discovering new talent and for promoting preexisting projects, like *What Is It?* In fact, as he talked, the two major issues facing YouTube—copyright and advertising—were brought into relief. While he didn't mind the clips that had been posted from studio films he'd acted in (Glover played George McFly, Michael J. Fox's father, in *Back to the Future*), he was concerned about the prospect of someone posting a bootleg of his own movie. "That's something I would be very litigious about," he said.

Turning to me, he asked, "What's the right word for the

conspiracy theories that Stevie has about viewership counts? Did she talk to you about that?"

"He's, like, 'Yeah, *right,*'" Ryan said.

"Who, me?" Glover asked. "No, no. I understand how those things happen, and it *does* have to do with corporate sensibilities. I don't know what's happening in her case—"

"Well, what do you *think* is happening?" Ryan interrupted.

"Well, if there would be things like that going on, the reason it would be happening is because they need to be figuring out how their sponsorship elements are going to be working, and corporate elements always get concerned about making audiences uncomfortable," Glover said. He brought up the fact that she plays a Hispanic character. "Racism is a hot issue, and if there's anything where people can be concerned about race issues, that's a sponsorship issue," he said. "In the 1970s, there was that Frito-Lay character—he sang a song, 'Ay, yii, yii, yiii, I am the Frito Bandito.' And at a certain point they had to get rid of that character, because he was thought to be making fun of the Hispanic community."

Glover was on a roll now, the wise 42-year-old actor-director schooling the naive 22-year-old in the ways of the world and warning her against the inevitable corruption of her utopian Internet democracy. "*My* movie definitely goes into those areas that corporate entities would be concerned about," he said. (Glover describes *What Is It?* as "the adventures of a young man whose principal interests are snails, salt, a pipe, and how to get home.") "And you have to start questioning the basis of the culture itself, being a free-market democracy, which has to do with capitalism—and there are *questions* about capitalism." Ryan's eyes began to glaze over. "You know, you can question Communism, you can question capitalism, you can question fascism," Glover went on. "They all have their innate evils, so to speak. And it's the corporate entity that ends up getting power within a capitalistic culture."

Ryan picked up on the doomsday narrative, if not the political theory. "Four months ago, when I was first on YouTube, it was *not* where it's at right now," she said. "I think Little Loca was, like, No. 5 most subscribed, and now, like, I'm No. 15—because why? There's all these other people they're featuring on there. And it's, like, bullshit."

A few weeks later, Ryan posted a new Little Loca installment. It begins with Loca walking down the street ("My dad let me take out the camera," she explains) and stumbling upon a vintage Bentley. "*Damn,* there's a bad-ass car up right here," she says. A man with long hair, funny glasses, platform shoes, and striped pants—Glover's Letterman outfit—approaches. "This fool be lookin' *scandalous* as *hell,*" Loca says. "Don't tell me that's *his* car. . . . Hey, wait a minute. You look like McFly, fool. Hey, are you Crispin Glover?"

"No," the man says, acting nervous.

"You're in a disguise," she says.

"No, I'm not," he says, now indignant. "People think that, and it's not true."

He drives off, but soon the scene cuts to Glover's front yard, where the tarps have been removed from the cars. "Look at these cars and this house," Loca says. "And you're going to try to say that you ain't Crispin Glover the *movie star?*"

"Movie star?" he asks. "*That* guy? That guy's an *idiot.*"

At about the four-minute mark, the screen fades to a "To be continued . . ." Then there is a printed advertisement: "See Crispin Hellion Glover in a live dramatic performance along with his feature film *What Is It?*" The film's trailer follows.

Less than 48 hours after Ryan uploaded the video she shot in front of me at her apartment, it was removed from the site, further fueling Ryan's suspicions. "They removed my video because YouTube always removes my videos," she wrote me in an e-mail. "It's OK, I still love them."

The real reason for her video's removal had nothing to do with any personal antipathy toward her among the YouTube staffers. YouTube had received a Digital Millennium Copyright Act complaint from a third party. Apparently, Ryan's mistake had been to edit her sketch too ambitiously, postdubbing the Wu-Tang Clan soundtrack that was distinct from the video recording and therefore digitally traceable. Had she merely played the song on her stereo while shooting the scene on the sofa, there would have been no way for anyone to detect it, short of watching every video on the site.

In September, Ryan found herself featured at last on the YouTube home page. But the video that YouTube had selected was one of the derivative silent films, a sentimental Chaplin tribute set to the piano music of Yann Tiersen. In just two weeks, it was viewed nearly 300,000 times—far more than any of the Loca videos. The response, judging from the 1,700 comments, was largely positive, although one viewer named Draftgon wrote, "Can anyone tell me why THIS video got on the front page? I don't see anything interesting about it." A girl named Morbidangel wrote, "WOW! Beautiful. . . . You made me cry."

That comment drew a reply from Ryan herself: "Awww, please don't cry. Art is beautiful, it reminds you that you're really not alone."

About the Contributors

Kevin Berger is the features editor at *Salon*.

Paul Boutin is a writer and editor who lives in San Francisco. He wrote "You Are What You Search" while working as a technical writer for search software maker Splunk. He is currently *Wired*'s managing editor for blogs.

Kiera Butler envies those who have made sorting through CD bins into a career, but she likes writing, too. Her work has appeared in *Columbia Journalism Review, Utne Reader,* and *OnEarth Magazine,* among other publications. She currently works as an associate editor at *Plenty* magazine.

When he's not tracking down interesting science and technology stories ("Say Hello to Stanley"), *Wired* contributing editor **Joshua Davis** spends his free time competing in the world's wildest competitions. It began when he entered the U.S. Armwrestling Nationals, where he lost every single match but ended up fourth (out of four) in the lightweight division. That earned him a spot on Team USA and sent him to Poland for the World Armwrestling Championships. *The Underdog,* his book about becoming an internationally ranked arm wrestler (as well as a sumo wrestler, matador, and backward runner), is now available from Random House. More info can be found at http://www.joshuadavis.net/.

Julian Dibbell has, in the course of over a decade of writing and publishing, established himself as one of digital culture's most thoughtful and accessible observers. He is the author of two books about online worlds—*Play Money: Or How I Quit My Day Job and Made Millions Trading Virtual Loot* (Basic, 2006) and *My Tiny Life: Crime and Passion in a Virtual World* (Henry Holt, 1999)—and has written essays and articles on hackers; computer viruses; online communities; encryption technologies; music pirates; and the heady cultural, political, and philosophical questions that tie these and other digital-age phenomena together. He lives in South Bend, Indiana.

Matt Gaffney's crossword puzzles have appeared in the *New York Times*, the *Los Angeles Times*, the *Wall Street Journal*, *Billboard*, and the *Onion*. His writing on puzzles and games has appeared in the *Washington Post*, *Slate*, and the *Weekly Standard*. He is the author of *Gridlock: Crossword Puzzles and the Mad Geniuses Who Create Them* and *The Idiot's Guide to Solving Crossword Puzzles and Other Word Games*.

Lori Gottlieb is a commentator for National Public Radio and has written for the *New York Times*, the *Los Angeles Times*, *Time*, *People*, *Elle*, *Glamour*, and the *I*, among many other publications. She is the author of the national best seller *Stick Figure: A Diary of My Former Self* and coauthor of *Inside the Cult of Kibu: And Other Tales of the Millennial Gold Rush* and *I Love You, Nice to Meet You*.

John Gruber writes and publishes Daring Fireball (http://daringfireball.net), a somewhat popular, strongly opinionated Weblog for Mac, Web, and user interface design nerds. He lives in Philadelphia with his wife and son.

Jeff Howe is a contributing editor at *Wired Magazine*, where he covers the entertainment industry, among other subjects. He is currently writing a book about crowdsourcing, a term he coined in his June 2006 *Wired* article. Before coming to *Wired* he was a senior editor at Inside.com and a writer at the *Village Voice*. In his 15 years as a journalist he has traveled around the world working on stories ranging from the impending water crisis in Central Asia to the implications of gene patenting. He has written for *Time Magazine*, *U.S. News & World Report*, the *Washington Post*, *Mother Jones*, and numerous other publications. He lives in Brooklyn with his wife, Alysia Abbott; their daughter, Annabel Rose; and a miniature black Lab named Clementine.

Kevin Kelly is Senior Maverick at *Wired Magazine* and author of *Out of Control* and *New Rules for the New Economy*. He is publisher of the popular Web sites Cool Tools and True Films and is on the board of the Long Now Foundation, which is building a 10,000-year clock.

Jaron Lanier is Interdisciplinary Scholar-in-Residence, Center for Entrepreneurship and Technology, University of California, Berkeley, and Scholar at Large, Live Labs, Microsoft Corporation. Lanier's interests include biomimetic information architectures, user interfaces, heterogeneous scientific simulations, advanced information systems for medicine, and computational approaches to the fundamentals of physics.

Preston Lerner is a contributing editor at *Popular Science* and a contributing writer at *Automobile Magazine.* He has written about technology (and other subjects) for the *New York Times Magazine,* the *Los Angeles Times Magazine, Wired, Smithsonian,* and *Air & Space.*

Farhad Manjoo is writing a book about how the Web, cable news, and talk radio contribute to a fragmented culture in which people indulge their own beliefs in favor of the truth. He is also a frequent contributor to *Salon.*

Justin McElroy is a journalist living with his wife, Sydnee, along the Ohio River in West Virginia. During the day, he's a mild-mannered newspaper reporter. But in the evenings (and on the occasional weekend) he has found the time to contribute to gaming publications such as the *Escapist, Joystiq, Computer Games Magazine, Gamezebo,* and *Gamers with Jobs: Press Pass.*

Ben McGrath is a staff writer at the *New Yorker.* He lives in Manhattan.

Katharine Mieszkowski is a senior writer for Salon.com, where she covers technology, health, and the environment. Her writing has appeared in *Rolling Stone, Glamour,* and the *Los Angeles Times.* Her commentaries have been featured on National Public Radio's *All Things Considered* and Public Radio International's *Living on Earth.* Katharine still hasn't spent the pennies she earned on Amazon Mechanical Turk.

Emily Nussbaum is an editor-at-large at *New York Magazine.* She lives in New York and writes frequently about pop culture and technology.

Jeffrey M. O'Brien is a senior editor at *Fortune*, covering the intersection of science, technology, culture, and business. He was a 2006 Templeton-Cambridge fellow in science and religion at the University of Cambridge.

Larry O'Brien is a software development consultant and contributing editor to *SD Times*. A well-known writer on software development topics, he blogs at http://www.knowing.net.

The ***Onion***, based in New York City, began as a student publication at the University of Wisconsin in 1988 and has since grown into America's preeminent satirical publication and what the *New Yorker* has called "arguably the most popular humor periodical in world history."

Adam L. Penenberg is a journalism professor at New York University and assistant director of the Business and Economic Reporting Program. In 1998, while a staff editor at Forbes.com, he garnered national attention for unmasking Stephen Glass as a fabulist, as portrayed in the 2003 film *Shattered Glass* (Steve Zahn plays Penenberg). His first book, *Spooked: Espionage in Corporate America* (Perseus Books, 2000), was excerpted in the *New York Times Sunday Magazine,* and his second, *Tragic Indifference: One Man's Battle with the Auto Industry over the Dangers of SUVs* (HarperBusiness, 2003), was optioned for the movies by Michael Douglas. A former columnist for *Slate* and *Wired News,* Adam is currently a contributing writer for *Fast Company* magazine.

John Seabrook has been a staff writer at the *New Yorker* since 1993. Prior to that, he was a staff writer at *Vanity Fair*. His work has also appeared in *Harpers, Vogue,* the *Nation,* and many other publications. He is the author of two books, *Deeper: My Two-Year Odyssey in Cyberspace* and *Nobrow: The Culture of Marketing, the Marketing of Culture*. A collection of his articles, entitled *Flash of Genius,* will appear in 2008. He lives in New York City with his wife and son.

Philip Smith has enjoyed writing all of his life. In 1994, he started writing and moderating for FireFly—one of the first social networking sites on the Internet, similar to today's

MySpace and FaceBook sites. In 1999, Smith started journaling on Slashdot. In 2002, he carried his writings to his own Web site, which eventually became his current blog on Blogger. In 2005, Smith's blog was listed in the Blog 100 by CNET.

Aaron Swartz is a writer, hacker, and activist from Cambridge, Massachusetts. He cofounded reddit.com, codeveloped the Markdown format, and coauthored the RSS 1.0 spec.

Clive Thompson writes about technology and culture for the *New York Times Magazine, Wired, New York Magazine,* and other publications. He also runs the tech-culture blog collision detection.net.

Jeffrey R. Young is a writer and editor for the *Chronicle of Higher Education.* After covering technology for the newspaper for more than 10 years, he recently began focusing on producing video and audio features for the Web. His freelance work has appeared in the *New York Times* and other publications.

Acknowledgments

Grateful acknowledgment is made to the following authors, publishers, and journals for permission to reprint previously published materials.

"The Artist as Mad Scientist" by Kevin Berger. First published in *Salon*, June 22, 2006. Reprinted with permission of the author.

"You Are What You Search" by Paul Boutin. First published in *Slate*, August 11, 2006. Reprinted with permission of the author.

"Listen to This" by Kiera Butler originally appeared in the May/June 2006 issue of *Columbia Journalism Review*. Reprinted with permission of *Columbia Journalism Review*.

"Say Hello to Stanley" by Joshua Davis. First published in *Wired Magazine*, January 2006. Reprinted with permission of the author.

"Dragon Slayers or Tax Evaders" by Julian Dibbell. First published in *Legal Affairs*, January/February 2006. Reprinted with permission of the author.

"The Ultimate Crossword Smackdown" by Matt Gaffney. First published in *Slate*, July 12, 2006. Reprinted with permission of the author.

"How Do I Love Thee?" by Lori Gottlieb. First published in *Atlantic Monthly*, March 2006. Reprinted with permission of the author.

"Good Journalism" by John Gruber. First published on http://dar ingfireball.net May 2, 2006. Reprinted with permission of the author.

"The Rise of Crowdsourcing" by Jeff Howe. First published in *Wired Magazine*, June 2006. Reprinted with permission of the author.

"Scan This Book" by Kevin Kelly. First published in *New York Times Magazine*, May 14, 2006. Reprinted with permission of the author.

"Digital Maoism" by Jaron Lanier. First published on *Edge* (www.edge.org) May 30, 2006. Reprinted with permission of *Edge*.

"This Is a Bike. Trust Us." by Preston Lerner. First published in *Los Angeles Times*, August 27, 2006. Reprinted with permission of the author.

Text design by Mary H. Sexton

Typesetting by Delmastype, Ann Arbor, Michigan

The text font is Granjon, which was designed in 1928
for Linotype by George Jones using Claude Garamond's
(1499–1561) late Texte (16 point) roman as his model.
It is named after the sixteenth-century French printer,
publisher, and lettercutter Robert Granjon (1513–89).
   —*Courtesy adobe.com and myfonts.com*